T0320762

SIGMA

A Knowledge-Based
Aerial Image
Understanding System

ADVANCES IN COMPUTER VISION AND MACHINE INTELLIGENCE

Series Editor: Martin D. Levine

McGill University
Montreal, Quebec, Canada

COMPUTER VISION FOR ELECTRONICS MANUFACTURING
L. F. Pau

SIGMA: A Knowledge-Based Aerial Image Understanding System
Takashi Matsuyama and Vincent Shang-Shouq Hwang

A Continuation Order Plan is available for this series. A continuation order will bring delivery of each new volume immediately upon publication. Volumes are billed only upon actual shipment. For further information please contact the publisher.

SIGMA
A Knowledge-Based Aerial Image Understanding System

TAKASHI MATSUYAMA

Okayama University
Tsushima, Okayama, Japan

and

VINCENT SHANG-SHOUQ HWANG

PLENUM PRESS • NEW YORK AND LONDON

Library of Congress Cataloging-in-Publication Data

Matsuyama, Takashi, 1951-
    SIGMA : a knowledge-based aerial image understanding system /
Takashi Matsuyama and Vincent Shang-Shouq Hwang.
       p.   cm. -- (Advances in computer vision and machine
intelligence)
    Includes bibliographical references.
    ISBN 0-306-43301-X
    1. Photographic interpretation--Data processing. 2. Image
processing--Digital techniques. 3. Expert systems (Computer
science) I. Hwang, Vincent Shang-Shouq. II. Title. III. Series.
TR810.M38 1990
006.3--dc20   89-29221
                                                             CIP

© 1990 Plenum Press, New York
A Division of Plenum Publishing Corporation
233 Spring Street, New York, N.Y. 10013

Printed in the United States of America

To Our Families

Preface

It has long been a dream to realize machines with flexible visual perception capability. Research on digital image processing by computers was initiated about 30 years ago, and since then a wide variety of image processing algorithms have been devised. Using such image processing algorithms and advanced hardware technologies, many practical machines with visual recognition capability have been implemented and are used in various fields: optical character readers and design chart readers in offices, position-sensing and inspection systems in factories, computer tomography and medical X-ray and microscope examination systems in hospitals, and so on. Although these machines are useful for specific tasks, their capabilities are limited. That is, they can analyze only simple images which are recorded under very carefully adjusted photographic conditions: objects to be recognized are isolated against a uniform background and under well-controlled artificial lighting.

In the late 1970s, many image understanding systems were developed to study the automatic interpretation of complex natural scenes. They introduced artificial intelligence techniques to represent the knowledge about scenes and to realize flexible control structures. The first author developed an automatic aerial photograph interpretation system based on the blackboard model (Naga1980). Although these systems could analyze fairly complex scenes, their capabilities were still limited; the types of recognizable objects were limited and various recognition errors occurred due to noise and the imperfection of segmentation algorithms.

In the 1980s, many vision researchers studied the computational aspects of visual information processing, such as methods for recovering three-dimensional information from two-dimensional images (e.g., stereo vision), and a few image understanding systems emphasizing so-called high-level vision were developed.

In artificial intelligence, the 1980s were a very fruitful period. Many practical expert systems were developed in various fields, including medical diagnosis, troubleshooting of electronic circuits, design of computer systems, and control of energy plants. Recently, many software tools to develop knowledge-based systems and new ideas have been proposed to cope with various problems encountered in the development of such practical systems.

We believe that this is an appropriate time to study extensively the problem of high-level vision:

1. Although many computational vision algorithms have been developed, all have advantages and disadvantages, so that a combination of them suggests a way to realize flexible and robust analysis. Such combination requires control by a high-level vision module.
2. Reasoning about time and geometric space and reasoning with uncertainty (or erroneous information) are crucial problems in achieving practical knowledge-based systems as well as image understanding systems. Thus the study of high-level vision contributes not only to image understanding but also to a wide spectrum of other research fields.
3. Recently neural networks have attracted many researchers in computer science. They can be considered as an alternative approach to the problem of intelligence: neurocomputing is based on pattern-directed (distributed) representation, while ordinary artificial intelligence uses symbolic representation. Since we can realize large-scale neural networks with modern LSI technologies, more capable machines may be created. Moreover, parallel computation and reasoning is an important problem to be investigated in all areas of computer science.

This book describes results of our joint studies on image understanding over several years. We designed the overall architecture of an image understanding system, SIGMA, and implemented a prototype system in 1983–1984 while the first author was at the University of Maryland as a visiting researcher. Later, the second author improved the prototype system to complete his Ph.D. thesis (Hwan1984a), and the first author studied knowledge-based image segmentation, which is one of the major research goals underlying SIGMA, at Kyoto University and Tohoku University, Japan.

Several papers on SIGMA have been published in academic journals and conference proceedings (Hwan1984b, Mats1985, Hwan1986, Mats1986, Mats1987). In writing this book, however, we did not want to simply bundle together the contents of these papers; instead we wanted to discuss various fundamental problems in image understanding and how we can cope with them by using modern artificial intelligence techniques. We have included surveys of aerial image understanding systems, spatial reasoning methods, and expert systems for image processing so as to clarify characteristics of SIGMA. In addition, we have discussed the theoretical foundations of our spatial reasoning in terms of the first-order predicate calculus. Descriptions in terms of formal logic enabled us to formulate the essential reasoning mechanisms used in SIGMA.

Major characteristics of SIGMA can be summarized as follows:

1. *Evidence accumulation for spatial reasoning.* A new flexible reasoning method based on spatial relations among objects is proposed. It integrates both bottom-up and top-down analyses into a unified reasoning process. It has much to do with nonmonotonic reasoning in artificial intelligence, which permits reasoning with incomplete knowledge.
2. *Distributed problem solving based on object-oriented knowledge representation.* In SIGMA, all recognized objects are regarded as active reasoning agents which perform reasoning about their surrounding environments based on their own knowledge. The system coordinates such local reasoning by independent reasoning agents to construct a globally consistent description of the scene. To model active reasoning agents, object-oriented knowledge representation is used.
3. *Knowledge-based image segmentation.* Many different image analysis processes can be used to extract features from an image. The knowledge-based image segmentation module in SIGMA reasons about the most effective image analysis process based on both knowledge about image processing techniques and the quality of the image under analysis. (We call such systems *expert systems for image processing.*) Guided by the knowledge, it performs trial-and-error image analysis automatically to extract features specified in a given goal.

It would be interesting to compare SIGMA with our former image understanding system based on the blackboard model (Naga1980); SIGMA emphasizes spatial reasoning and top-down goal-directed image segmentation, while the former system incorporated sophisticated image analysis procedures to recognize objects based on their various spectral (i.e., color) and spatial properties.

The system as implemented analyzes black-and-white aerial photographs of suburban housing developments and locates cultural structures such as houses, roads, and driveways. Thus, the book will be interesting and valuable to those who are engaged in remote sensing as well as in image processing, image understanding, and artificial intelligence. We hope that this book contributes to the further development of these research areas.

Takashi Matsuyama
Vincent Shang-Shouq Hwang

Okayama and McLean

Acknowledgments

We are grateful to Professors Azriel Rosenfeld and Larry Davis of the University of Maryland for their support of this research and for constructive comments on technical issues. The first author also thanks Professor Makoto Nagao of Kyoto University and Professor Takayasu Ito of Tohoku University for giving him this research opportunity and for their continual encouragement and advice.

Since this research was jointly performed at the Computer Vision Laboratory of the University of Maryland, the Department of Electrical Engineering of Kyoto University, and the Department of Information Engineering of Tohoku University, many students and members of the technical staffs of these universities helped us to implement the system and conduct experiments. We would like to thank all of them for their assistance.

Contents

SIGMA

Chapter 1

Introduction

The term *image understanding* has been widely used since work on image understanding in the United States started in 1975. It refers to knowledge-based interpretation of visual scenes by computers and consequently denotes an interdisciplinary research area including signal processing, statistical and syntactic pattern recognition, artificial intelligence, psychology, and even neurophysiology. In the early literature, it was called *scene analysis*. *Computer vision* is also widely used to refer to a similar research area; but while computer vision emphasizes computational aspects of visual information processing, such as measurement of three-dimensional shape information by visual sensors, image understanding stresses knowledge representation and reasoning methods for scene interpretation. Although understanding time-varying scenes is a very important topic in image understanding and computer vision, in this book we confine ourselves to image understanding of static scenes.

In this chapter, we first describe, as an introduction, a general framework for image understanding in which the importance of discriminating levels of information is emphasized. Then, we provide an overview of the image understanding systems for aerial photograph interpretation that have been developed so far and discuss their limitations. Three general problems in image understanding are noted: (1) unreliable segmentation, (2) representation of and reasoning based on geometric information, and (3) reasoning with incomplete information.

This discussion is followed by surveys of various approaches to spatial reasoning and of control structures in image understanding. In the last section, we outline our fundamental ideas to solve the above problems, based on which we designed our image understanding system, SIGMA.

1.1. FRAMEWORK FOR IMAGE UNDERSTANDING

A primary objective of image understanding systems (IUSs) is to construct a symbolic description of the scene depicted in an image. In contrast, image processing transforms one image into another and pattern recognition classifies and labels objects represented by feature vectors. IUSs analyze an image or images to *interpret* the scene in terms of the object models given to the IUSs as knowledge about the world. Here *interpretation* refers to the correspondence (i.e., mapping) between the description of the scene and the structure of the image (Fig. 1.1). It associates objects in the scene (e.g., houses, roads) with *image features* in the image (e.g., points, lines, regions). Once the description of the scene has been constructed, computer systems can answer various queries about the scene (e.g., "How many houses exist in the scene?") and can perform physical operations by controlling robot manipulators (e.g., pick up and move physical objects). It is in this sense that we can say IUSs *understand* the scene.

IUSs consider a scene as a heterogeneous structure composed of mutually related objects of different kinds, while object detection (recognition) systems dichotomize the world into the *target object* and the *background*. Since the objective of the latter systems is to extract and recognize specific target objects, most of the knowledge used is confined to that about their intrinsic properties (e.g., color, shape) and little knowledge about the entire world is used. However, this dichotomy is valid only in *artificial* environments, for example, an isolated industrial part on a conveyer of uniform color. In reality, when we look around, we always see the scene jumbled with various objects: tables, chairs, lighting devices, coffee cups, and so on.

Most *natural* scenes are composed of objects of various kinds which are related to each other through their *functions*: coffee cups are on tables, chairs are near tables, table tops are parallel to the floor, and so on. Thus, to understand the scene, we need knowledge about (spatial) relations between objects as well as knowledge about their intrinsic properties. Since some objects may be occluded by others and only partial information about mutual relations plays an especially important role in understanding natural scenes. Using knowledge about relations,

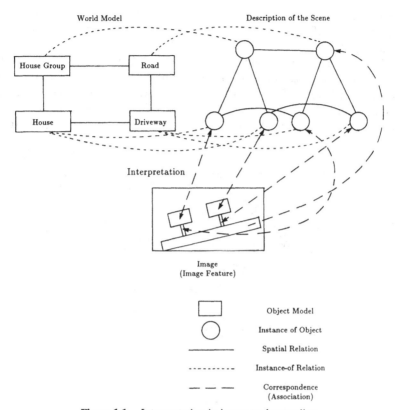

World Model   Description of the Scene

Interpretation

Image
(Image Feature)

Object Model

Instance of Object

Spatial Relation

Instance-of Relation

Correspondence
(Association)

Figure 1.1.   Interpretation in image understanding.

IUSs conduct reasoning about the structure of the scene. This clearly discriminates image understanding from statistical pattern recognition, in which all processing is done based on *attributes* of objects. The knowledge provided to the system, including information on the intrinsic properties of and mutual relations between objects in the world, is called the *world model* (Fig. 1.1).

In image understanding, discrimination between information about the scene and that about the image is crucial. The scene is described in terms of objects in the world, while the structure of the image is described in terms of image features. There exists a large gap between these two levels of information: usually the scene is three-dimensional (3D) while the image is two-dimensional (2D). The task of IUSs is to fill in the gap by computation and reasoning and to construct an interpretation (i.e., an association between the information at these two levels). A major reason for the limited performance of early IUSs (Barr1971, Prep1972, Yaki1973, Tene1977) is that they tried to jump over this gap in a single step.

In modern IUSs, the information is organized at several different levels. Many analysis processes are incorporated to analyze the information at each level and to transform the information from one level to another. Figure 1.2 illustrates the levels of information used in IUSs (see also Fig. 1.1) and the analysis processes which transform the information

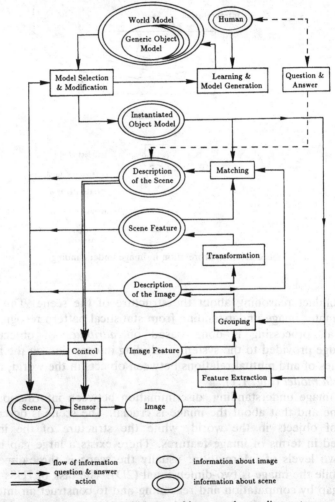

Figure 1.2. General architecture of image understanding systems.

from one level to another. The vertical sequence of ellipses in Fig. 1.2 represents the levels of information used in IUSs. The information about the image is represented by single ellipses, while that about the scene is symbolized by double ellipses. The analysis processes are illustrated by rectangles, and the flow of information is represented by arrows connecting ellipses and rectangles. In what follows, we summarize the levels of information, the analysis processes, and the knowledge used for the analysis based on Fig. 1.2.

1.1.1. Information about the Image

First, the physical scene is sensed by one or more sensors to produce one or more images. There are many different types of sensors: ordinary visual sensors like TV cameras, multispectral scanners, range sensors, and so on. Each type of sensor measures a different type of information about the scene: brightness, color, depth, and so on. Among these, the depth information measured by range sensors is not what we would call information about the image but rather information about the scene, although it is usually represented by an image (i.e., iconic description). In other words, we can directly obtain scene features by analyzing range images. Note that lighting devices should also be considered as important parts of sensor systems; by controlling lighting, we can measure various information about the scene [e.g., the use of photometric stereo for measuring the surface orientation of 3D objects (Wood1978)]. Although the question of how to design sensor systems is an important problem in image understanding, it is beyond the scope of this book (see Ball1982 for various visual and range sensors and lighting devices used in computer vision). Here, we consider images taken by ordinary visual sensors like TV cameras. [We sometimes use the term *pictures* instead of *images* (e.g., gray pictures).]

A raw input image at the signal level is analyzed to produce primitive *image features,* such as points, lines, and regions, which then are organized (grouped) into more informative image features, for example linking small disjoint line segments into a continuous line and merging irregularly shaped small regions into a compact region. Mutual geometric relations between image features are established to produce the structural description of the image. These processes are called *segmentation.* In Fig. 1.2, Image → (Feature Extraction) → Image Feature → (Grouping) → Description of the Image corresponds to segmentation.

Currently, it is widely accepted that this structurization of the information about the image should be done without using semantic knowledge about a specific scene: the rule "since the color of grasses can be green or yellow, merge green regions with neighboring yellow regions" incorporates domain-specific knowledge into segmentation (Yaki1973). Major sources of the knowledge used for segmentation are *image models* to extract primitive image features and so-called *Gestalt laws* to group them.

As noted above, various different types of information are measured in the form of images to facilitate the interpretation of visual scenes: range images representing the depth of the scene, and images obtained by touch sensors, infrared (temperature) sensors, and radars. Although all are represented as 2D images, their characteristics are very different. Thus we need to use different image models in order to process them to extract meaningful information. The ordinary heuristic, "regions are of uniform color and their boundaries are step edges," is too simple to model images: the brightness of a curved surface changes gradually, and a 3D edge between a pair of planar surfaces is depicted as a roof edge in a range image. Haralick (Hara1981) proposed a functional approximation method to extract image features. He approximated the spatial distribution of pixel values in a local area by a polynomial function and defined various image features based on the shape of the function.

Marr (Marr1975, Marr1982) proposed edge-based image analysis using Laplacian operators of different resolutions and called the edge-based symbolic description of the image the *primal sketch*. He formulated several grouping methods using the primal sketch. Zucker (Zuck1975) called domain-independent knowledge about grouping *general-purpose models*. Lowe (Lowe1985a,b) emphasized the importance of perceptual organization (grouping) in image understanding.

1.1.2. Transforming the Information from the Image Domain to the Scene Domain

An image sensor projects a (3D) scene onto a (2D) image, so that a role of IUSs is to *back-project* the information in the image domain to that in the scene domain. Image features are mapped into scene features by this transformation process, and correspondences between features at these two levels are recorded. In Fig. 1.2, analysis process Transformation performs this back-projection and establishes the correspondences.

It should be noted that scene features obtained by this transformation correspond to those parts of object surfaces in the scene which are observable from the sensor (camera). Marr (Marr1982) called such information $2\frac{1}{2}$ description. It refers to partial 3D information about observable object surfaces.

The knowledge required for this back-projection is knowledge about physical properties of the sensor and the geometry of the original projection (photographing conditions). Extensive studies have been conducted since the early 1980s to recover 3D information from various 2D image features: shape from shading, texture, and contour; stereo vision; structure from motion. Many algorithms based on photogeometry have been proposed for these problems in *computational vision* (Ball1982, Marr1982). We will not discuss this topic further, since it is beyond the scope of this book.

1.1.3. Information about the Scene

The information about the scene can be classified into several types depending on levels of abstraction. The most abstract information is the world model. It includes models of *generic* objects and their mutual relations (see Fig. 1.2). Usually each generic object model is represented by a structural description including variables. The utilization of variables allows the representation of abstract objects (see Fig. 1.4 for example).

The recognition of objects in image understanding is realized by matching object models with features extracted from the image. By this matching process, abstract object models are *instantiated* to generate specific concrete models: a numerical value is assigned to and/or constraints on allowable values are associated with each variable in an abstract object model. Instantiated object models refer to observed practical objects in terms of which the description of the scene is constructed. We call them *object instances* (Fig. 1.1).

Since matching is the kernel process in IUSs, it can be very complicated to realize flexible recognition capability.

First, the matching process can be divided into two subprocesses: model selection and the process of establishing correspondences. Since various object models are included in the world model, we first have to select an appropriate one to match with the observed features. In Fig. 1.2, analysis process Model Selection & Modification performs this model selection to generate Instantiated Object Model. Various levels of information can be used for the model selection: Description of the Image, Scene Feature, and (partial) Description of the Scene. (In Fig. 1.2, the incoming arrows at Model Selection & Modification represent the information flow from these types of information.) On the other hand, analysis process Matching in Fig. 1.2 establishes precise correspondences between the selected object model and observed features to generate Description of the Scene. Exact locations of recognized object instances and spatial relations to others are determined by this process.

Second, there are two schemes of matching: (1) Transform image features into scene features, which then are matched with object models (Barr1978). (2) Match image features directly with object models. In the latter scheme, the matching process involves the determination of the parameters of the geometric transformation between the image and the scene (Robe1963). Lowe (Lowe1985b) stressed the utility of this matching scheme when a concrete geometric object model is given *a priori*. Ballard (Ball1981) proposed the generalized Hough transformation to determine the geometric transformation between an object model and an image.

Finally, if some parts of the instantiated object model do not match any observed features, the matching process may activate segmentation again to extract new features corresponding to such missing parts. We call this process *top-down analysis,* while the analysis process so far described (i.e., segmentation followed by transformation and matching) may be termed *bottom-up analysis.* In top-down analysis, the information about instantiated object models is used to control segmentation. In Fig. 1.2, this information flow is illustrated by downward arrows to Grouping and Feature Extraction. We will discuss characteristics of bottom-up and top-down analyses in more detail in Section 1.5.

Since the description of the scene is *constructed* from object instances, it includes information which is *not* contained in the input image at all; for example, complete 3D shapes of objects are included in the description, even if some parts of them (e.g., back sides) are occluded and cannot be seen in the image. Thus the matching process in IUSs does not simply classify given image features into object categories but *constructs* a complete description of the scene even if it is only partially depicted in the image. The world model complements the necessary (missing) information to construct the full scene description.

This view of visual perception as an active constructive process has been our fundamental principle in conducting the research reported in this book. Gregory (Greg1970) demonstrated varieties of psychological evidence supporting this principle. In Section 1.3 we will discuss why we need the world model to construct the scene description, and we will propose our scheme of *constructive reasoning* in Sections 2.2 and 2.3.

The constructed scene description is used to answer queries about the scene, to control a sensor to obtain more information about the scene, and to manipulate physical objects in the scene. Figure 1.2 includes these additional processes using the scene description.

1.1.4. Types of Knowledge Used in IUSs

As described above, IUSs require a wide spectrum of knowledge to understand the image. We can classify it into the following three types:

1. *Scene domain knowledge.* This type of knowledge includes intrinsic properties of and mutual relations among objects in the world. It is described in terms of the terminology defined in the world: names of objects and their constituent parts, the geometric coordinate system to specify locations and spatial relations, the physical scale system to measure various size properties (i.e., length in *meter*, area in *meter*^{2}, volume in *meter*^{3}), and so on.
2. *Image domain knowledge.* This type of knowledge is used to extract image features from an image and to group them to construct the structural description of the image. It is described in terms of the terminology defined in the image domain. This terminology must not be confused with that for describing the scene: a word "adjacent" in the scene domain knowledge must be clearly discriminated from that in the image domain knowledge (Reit1987).
3. *Knowledge about the mapping between the scene and the image.* This type of knowledge is used to transform image features into scene features and vice versa. That is, it defines translation rules between the two terminologies used to describe the scene and the image domain knowledge. Knowledge about the photogeometry is a typical example: the focal length of the camera, aspect ratio, resolution, and so on. In measuring colors and multispectral information, knowledge of the spectral characteristics of optical filters is important to calibrate measured values to obtain the physical energy in different spectral channels.

Besides these types of knowledge, IUSs often need *control knowledge* to guide reasoning processes: rules to determine what objects should be detected first, where to analyze next, and so forth. Since a wide spectrum of object models are included in the world model, IUSs have to conduct searching in a huge problem space to identify the objects depicted in the image: an image feature can be interpreted as parts of many different objects, so that IUSs must examine each possible interpretation. The control knowledge includes heuristics to prune the search space, a step which is crucial in realizing efficient reasoning (searching). Using *metarules* representing such control knowledge, we can realize IUSs with efficient and flexible control structures. (Examples of metarules will be given in Sections 3.7 and 4.1.6.)

1.2. AERIAL IMAGE UNDERSTANDING SYSTEMS: AN OVERVIEW

Automatic interpretation of aerial photographs has long been a dream in remote sensing, and many methods have been developed for analyzing remotely sensed imagery:

1. Statistical classification methods based on multispectral properties of pixels.
2. Target (object) detection by template matching. These first two methods were developed at an early stage of the research. Currently, the former methods are widely used for pixel-based image analysis, and the latter for detecting fixed-shape objects.
3. Shape and texture analysis by using picture processing techniques. To analyze structures of complex aerial photographs, we have to use spatial information as well as spectral information. For example, edge detection and shape analysis are useful to recognize irregularly shaped objects such as roads, and texture analysis to analyze forest areas. Picture processing techniques have allowed the recognition of objects based on their spatial properties.
4. Use of structural and contextual information. Many objects have complex internal structures and the recognition of objects is sometimes affected by their surrounding environments. For example, an airport consists of runways, taxiways, and terminal buildings. A bridge should be recognized as a part of a road overpassing a river or another road. To recognize such objects, knowledge about structures of objects and their environments (contexts) is required.
5. Image understanding of aerial photographs. To automate the interpretation process for complex aerial photographs, we must incorporate all the analysis methods described above. In particular, in interpreting complex scenes of urban and suburban areas, the knowledge about objects and the control knowledge become crucial. Since knowledge representation and reasoning have been major research topics in artificial intelligence (AI), it is natural to draw upon AI techniques to develop image understanding systems for aerial photograph interpretation. Here we call such systems *aerial image understanding systems*.

The problem of aerial image understanding has been attracting many researchers. One reason for this is of course its practical utility: many tasks of monitoring natural and artificial environments can be automated with aerial image understanding systems. Another reason is that aerial photographs are so complex that simple analysis methods do not work well and sophisticated computation, reasoning, and control mechanisms are required to analyze them. That is, aerial image understanding can be considered as a good testbed to investigate the feasibility of general image understanding. In particular, while computational methods have been extensively studied in computer vision for robotics and industrial applications, knowledge representation and reasoning schemes have been major research topics in aerial image understanding. Thus many knowledge-based systems have been developed for aerial image under-standing (Binf1982).

In the rest of this section, we briefly survey several aerial image understanding systems developed so far and discuss their limitations and problems.

Object detection and recognition is an important task in image understanding. Using various spatial and spectral properties of objects, we can develop object detection experts (specialized analysis modules). In the late 1970s, many such experts were developed to recognize specific types of objects in aerial imagery. For example, the road expert developed at SRI (Quam1978, Agin1979) traced complex highways in aerial photographs and extracted vehicles on them. Although these experts incorporate knowledge about objects, they are too simple to be considered as aerial image understanding systems; they are just special-ized image analysis programs in which specific object models are encoded.

Since an aerial photograph contains many different types of objects to be recognized, we have to develop various object detection experts and integrate them into a unified system. The problem is how to integrate such experts in a flexible way. The first author developed an aerial image understanding system based on the *blackboard model* (Naga1980). Figure 1.3 illustrates the configuration of the system. The blackboard is the database in which all the information about properties of and relations among regions and recognized objects is stored. A group of analysis modules interface with it in a uniform way to examine data for the analysis and to return their analysis results. Since all recognition results from object detection subsystems are written on the blackboard, each subsystem can use them to detect new objects. For example, once roads have been detected by the road detection subsystem, the car detection subsystem can use that result to select candidates for cars: small regions

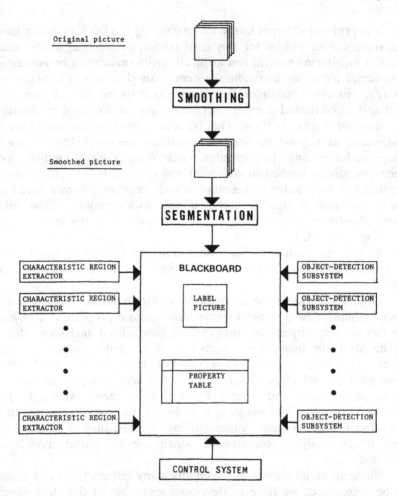

Figure 1.3.   Image understanding system based on the blackboard model (from Naga1980).

in the recognized roads are possible candidates for cars. The system controller always monitors the content of the blackboard and (1) resolves conflicting analysis results obtained independently by multiple object detection subsystems, and (2) corrects errors of the initial region segmentation based on the analysis results from object detection subsystems.

This system can analyze fairly complex aerial photographs. The success of this system can be ascribed to the modular system organization of the blackboard model and the flexible control structure as well as the sophisticated image analysis procedures used by the object detection systems.

From an image understanding point of view, however, its capability is still limited:

1. All knowledge about objects is encoded in the programs (object detection subsystems), so that it is not clear what knowledge is used and modification of the knowledge requires complicated programming.
2. All of the analysis is done based on 2D object appearance. The distinction between the image-level and the scene-level information is not so clear.
3. Most of the knowledge used for object recognition is knowledge about attributes of objects. Little knowledge about the internal structures of and spatial relations among objects is used.

ACRONYM (Broo1981) is a general image understanding system for detecting complex 3D objects. It demonstrated how AI techniques can be used to solve the above problems:

1. *Symbolic knowledge representation.* In ACRONYM, models of objects to be recognized are symbolically described in terms of *frames* (Mins1975). A frame is a data structure to represent structured knowledge about an object. It consists of a set of *slots,* in which various types of information, such as attributes of the object, relations to other objects, and procedures to compute properties, are stored. The concept of a frame has been widely used in many IUSs: MAPSEE (Have1983) for map interpretation, VISIONS (Parm1980) for outdoor scene interpretation, and our aerial image understanding system, SIGMA. (See Sections 2.4 and 4.2 for the knowledge representation methods used in SIGMA.) Symbolic knowledge representation is very important for general image understanding systems; the same system can be used to analyze various different scenes by changing the world model.

2. *3D object model.* ACRONYM uses *generalized cylinders* (Ball1982, Shaf1983) to represent models of 3D objects. A generalized cylinder is described by (a) a spine (axis): a 3D curve, (b) a cross section: a planar region, and (c) a sweeping rule to transform the cross section as it is moved along the spine. A generalized cylinder is defined by the 3D subspace swept by the cross section when it is moved along the spine. During the sweep, the shape of the cross section is changed according to the sweeping rule. Generalized cylinders have often been used to model shapes of 3D objects in image understanding; 3D objects with axes can

```
a   (define cone FCONE having
        main-cone (define simple-cone having
                      cross-section (define cross-section having
                                        type      circle
                                        radius    FUSELAGE-RADIUS)
                      spine          (define spine having
                                        type      straight
                                        length    FUSELAGE-LENGTH)
                      sweeping-rule  constant-sweeping-rule))


b           frame type   frame name
        (define object JET-AIRCRAFT having
            ┌ subpart           FUSELAGE
            │ subpart           PORT-WING
            └ subpart           STARBOARD-WING)

        ─ slots to represent part names

        (define object FUSELAGE having
            cone-descriptor FCONE
            ┌ subpart           PORT-STABILIZER
            ├─▶ subpart         STARTBOARD-STABILIZER
            └ subpart           RUDDER)


c   (constraint B747 with
            (c_interval FUSELAGE-LENGTH 50.0 70.0)
            (=:FUSELAGE-RADIUS 3.25)
            (=:ENGINE-RADIUS   1.5)
            (=:ENGINE-LENGTH   6.3)
            (=:WING-LENGTH     31.89))
```

Figure 1.4. Symbolic model representation of 3D objects (from Broo1981). (a) Shape description by a generalized cylinder. (b) Structural model representation using PART-OF relations. (c) Constraints for B747.

be naturally described in terms of generalized cylinders (Marr1982). Figure 1.4a shows a frame representing a part of an airplane in terms of a generalized cylinder. Explicit 3D models allow the recognition of objects from any viewpoint.

3. *Structural knowledge organization.* The world model in ACRONYM is organized into two types of hierarchies: the composition hierarchy based on *PART-OF* relations and the generalization/specialization hierarchy based on *A-KIND-OF* relations. The former is used to represent geometric structures of objects and the latter to organize objects based on levels of abstraction. Figure 1.4b shows frames representing the structure of a generic (abstract) model of airplanes. Each spatial relation between a whole (e.g., JET-AIRCRAFT) and a part (e.g., FUSELAGE) is described in another frame, in which exact geometric relations such as 3D distances and angles are stored. A specific model like B747 is represented by specializing the generic model. The specialization is represented by associating a set of constraints with the generic model. Figure 1.4c shows a frame describing the constraints for a B747.

Although ACRONYM demonstrated a framework for general 3D

image understanding, its recognition results are not so satisfactory. The major reasons for this are that (1) The segmentation process to extract image features is not perfect. (2) The recognition strategy is purely *bottom-up*: extract features from an image and then match them with models. *Top-down* analysis to detect missing parts of objects should be incorporated.

Selfridge (Self1982) developed an aerial image understanding system with top-down image analysis capability: a group of image analysis procedures are activated during the analysis to detect missing objects. First, using the models of already recognized objects, expectations about locations of missing objects are generated. Then the image analysis procedures are applied to the expected locations to extract image features corresponding to the missing objects. Extensive discussions on bottom-up and top-down image analyses will be given in Section 1.5.

Herman (Herm1986) developed a 3D aerial image understanding system for urban areas. It analyzes a set of stereo aerial photographs to construct 3D descriptions of buildings in downtown Washington, D.C. His fundamental idea is as follows. Since by analyzing a single pair of stereo images we can obtain only partial 3D shape information about buildings, the system should *incrementally* construct the scene description by using a set of stereo pairs taken from different viewpoints. Analyzing a pair of stereo images, it first constructs a *complete* scene description which seems plausible in light of the observed information. Since the constructed description includes many hypothetical objects which have not been verified by the observed data, the system records mutual dependencies among verified and hypothetical objects and prepares for future modifications of the description: which verified objects support the existence of which hypothetical objects. When a new pair of stereo images is given, the system matches the old description with the newly observed information. Then, the system constructs a new description by integrating all the information so far obtained. This system uses a *truth maintenance system* (Doyl1979) as its module to record the dependency and to manage hypothetical objects.

McKeown (McKe1985) developed a rule-based aerial image understanding system, SPAM, for airport scenes. It contains the image/map database MAPS (McKe1983) as its knowledge source. In SPAM, all extracted image features are first transformed into the map space, in which all reasoning for object recognition is performed. Several different types of *production rules* are used for the different tasks required for interpretation. A production rule is described by

IF (precondition) THEN (action)

When the precondition is satisfied, the action is activated. The utility of production rules in representing knowledge has been widely recognized in various expert systems (Haye1983). In SPAM, the system first applies all rules for object recognition independently to generate partial interpretations, and then a set of rules for consistency examination are applied to examine the consistency among the generated interpretations and to evaluate their reliabilities. In this style of reasoning, the system maintains multiple different possible interpretations simultaneously, since it cannot determine a unique interpretation based on partial information included in an image.

Herman's system *believes* a single interpretation at a time and changes the belief if inconsistency is detected, while SPAM maintains all possible interpretations and selects one at the final stage of the analysis. (ACRONYM uses the latter method to maintain interpretations.) The former method is based on Doyle's truth maintenance system (Doyl 1979) and the latter is similar to de Kleer's assumption-based truth maintenance system (de K1986). Both truth maintenance systems were designed to realize *nonmonotonic reasoning* (Spec1980), which is one of the recent central problems in AI.

Through experiments with these aerial image understanding systems, the following points become clear:

1. Various AI techniques are effective in image understanding: the blackboard model, symbolic knowledge representation by frames and production rules, and hierarchical knowledge organization based on PART-OF and A KIND-OF relations.
2. The use of explicit 3D object models allows the interpretation of fairly complex 3D scenes.
3. IUSs need sophisticated control structures to realize flexible analysis by incorporating both bottom-up and top-down analyses.
4. Image understanding has much to do with recent AI problems such as truth maintenance and nonmonotonic reasoning.

However, there still remain many problems to be solved in image understanding. We will discuss them in the next section.

1.3. PROBLEMS IN IMAGE UNDERSTANDING

In general, IUSs need to perform two types of tasks: (1) *segmentation*: extract meaningful image features from an image to construct the description of the image; and (2) *interpretation*: establish the correspondence between image features and object models to construct the description of the scene.

Segmentation refers to processing at the signal level. It can be realized by using various image processing techniques developed so far (Rose1982): noise filtering, edge detection, region segmentation, texture analysis, and so on. On the other hand, the process of interpretation involves diverse types of processing (some are numeric and others symbolic): mathematical (probabilistic) analysis for classification, graph theoretic analysis for matching, search processes for object identification, and so on. We need to integrate all these analysis methods to realize versatile IUSs.

We have identified three problems in image understanding which have not yet been treated successfully.

1.3.1. Unreliable Segmentation

Although many methods of segmenting images to extract points, lines, and regions have been developed, none of them is perfect: erroneous image features are extracted and some meaningful ones cannot be detected. There are several reasons for this:

1. Although most segmentation methods perform processing uniformly over the entire area of an image, its quality (S/N ratio, contrast) often varies depending on location: shading by uneven illumination, highlights on reflective surfaces, and shadows are ubiquitous in natural scenes. As a result some portions of the image can be analyzed correctly while others are not.

2. As noted before, grouping primitive image features is an important process in constructing the description of the image. Although many people have developed various computational methods of grouping (Marr1975, Zuck1975, Lowe1985a), their capabilities are still limited.

(More practical problems in developing segmentation procedures will be discussed in Section 4.1.1.)

One idea to cope with problem (1) above is to incorporate multistage plan-guided analysis: first generate a plan representing the global rough structure of the image and then analyze local areas guided by the plan (Kell1971). In plan-guided analysis, adaptive image processing based on local image quality can be realized, a strategy which improves the efficiency and reliability of segmentation. Analysis based on the pyramid data structure (Tani1975) and scale space filtering (Witk1983) are examples of this approach using multiple images/operators with different resolutions. Many IUSs incorporate plan-guided analysis to realize efficient object recognition (Hans1978, Rose1978, Naga1980, Ohta1980).

As for problem (2), given a model of image features to be extracted, we can use the Hough transform (Duda1973, Ball1981) to group primitive image features to form meaningful ones. Although it is computationally expensive, it is insensitive to noise and irregularities (e.g., gaps between line segments) which often disturb the grouping process.

The imperfection of segmentation leads to the following two problems: (1) introduction of erroneous information (i.e., erroneous image features) and (2) lack of meaningful information (i.e., missing image features). Since these problems are inevitable,* IUSs should incorporate versatile mechanisms to cope with them: identification and elimination of erroneous information, and investigation and extraction of missing information. As discussed above, if we are given models (plans) of images and/or image features, we will be able to realize reliable image segmentation. Thus, the problem is how we can provide appropriate models to the segmentation process in IUSs; the models should be adopted to the quality of the image and the target image features to be extracted.

1.3.2. Representation and Reasoning Based on Geometric Information

In image understanding, how to represent and use geometric information is a major problem. Although many methods of shape analysis have been developed to characterize shape features in image processing (Rose1982), most of them are confined to 2D image features. In computer vision, computer graphics, and computer-aided design, various 3D models have been proposed to represent 3D shapes of objects: edge-based, surface-based, and volumetric models (Ball1982).

---

* It would be arguable whether or not perfect segmentation can be realized. We believe that it is almost impossible to extract all meaningful image features perfectly at the initial segmentation: (1) meaningfulness can only be defined in light of some global context and the objective of the analysis, and (2) nonuniform image quality and noise are ubiquitous.

Thus it would be reasonable to first establish global context and estimate image quality at each location based on the result of imperfect segmentation and then to recover the imperfection. From a more fundamental point of view, it should be noted that, although segmentation followed by recognition is the standard process in current IUSs, Fukushima (Fuku1988) pointed out that segmentation is performed *after* recognition in his neutral network system.

All of them have advantages and disadvantages, so that we have to select appropriate models depending on the task at hand. In image understanding, we have to use those models by which the matching process with image/scene features can be facilitated. Several *appearance-based* 3D object models have been proposed for this purpose (Mins1975, Koen1979, Feke1984, Ikeu1987).

In addition to these shape feature and geometric models, IUSs need to use spatial relations among objects in constructing the scene description. In the recognition of manmade objects, knowledge about their spatial relations is especially important; their *functions* are often represented by their geometric properties and the spatial relations among them. For example, the length of a runway limits the types of airplanes that may be found in the airport. How to represent such functions in terms of spatial relations and how to use them for object recognition are major problems in image understanding.

We call reasoning based on spatial relations *spatial reasoning* and will discuss various schemes for spatial reasoning in the next section.

1.3.3. Reasoning with Incomplete Information

Recently it has often been noted that the problem of image understanding is *underconstrained*; the information included in input images is limited and is not sufficient to construct (recover) the complete description of the scene. In other words, straightforward mathematical formulations of image understanding problems lead to underconstrained equations with infinite solutions. For example, a straight line in an image can be interpreted as any member of the infinite set of 3D curves whose projections on the image plane coincide with that straight line.

In practical IUSs, this *intrinsic insufficiency* of input information is corrupted by the imperfection of segmentation, which, as noted in Section 1.3.1, leads to the introduction of erroneous information and a lack of meaningful information. We may call the insufficiency caused by segmentation *artificial insufficiency*. Thus IUSs need to perform reasoning with *incomplete* information: insufficient information, including errors. In such reasoning, we must first of all complement the imperfection of segmentation (i.e., elimination of errors and extraction of missing information). Although it is possible to recover the artificial insufficiency, IUSs in principle cannot recover the intrinsic insufficiency. Then what IUSs can do best is to construct *plausible* descriptions of the scene based on the available information; since the amount of input information is limited, multiple descriptions may be constructed. In short, IUSs perform reasoning to construct plausible scene descriptions while complementing the imperfection of segmentation.

In AI, reasoning with incomplete knowledge is considered a major characteristic of human reasoning, and many nonmonotonic reasoning schemes have been proposed (Spec1980). In nonmonotonic reasoning, various deductive reasoning methods are used to derive plausible conclusions in the light of available knowledge. In such reasoning, how to define the plausibility is the kernel problem: the logical consistency and minimization of logic models are usually used to define the plausibility in nonmonotonic reasoning (Gene1987). (In Section 2.3 we will discuss the logical foundations of spatial reasoning by SIGMA in the context of nonmonotonic reasoning.)

The following are four major approaches to cope with incomplete information in image understanding:

1. *Use of multiple sensors.* Use multiple (different) sensors to increase the amount of input information: stereo vision, time-varying images, use of color, and multispectral, range, and touch sensors, for example.
2. *Optimization.* Consider the information obtained from an input image as constraints and optimize the evaluation function which measures the goodness of match between image features and object models.
3. *Probabilistic reasoning.* Associate a probability with each piece of information and measure the degree of match based on the probability. The scene description with the highest probability is selected as the correct (i.e., most plausible) interpretation.
4. *Knowledge-based analysis.* In addition to pure geometric models, incorporate rich control knowledge to guide the matching process and to construct the description of the scene. Image features which do not match the models are considered as errors and many hypothetical features are generated based on the models to complement the insufficient input information.

The first approach is widely used to improve the reliability of data and the accuracy of the analysis in 3D object recognition: trinocular stereo, color cameras, and range sensors are all employed. In complex scenes, however, we cannot obtain complete information about objects no matter how many sensors we use (the intrinsic insufficiency of input information): usually the bottom faces of objects are touching other objects supporting them and some portions of objects may be occluded by other objects. The coordination of vision and robot systems is required to obtain information about such occluded surfaces. From a practical point of view, the complexity and cost of sensor systems limit the number of sensors, and the increased size of the observed signal data raises the computation cost. Moreover, additional (difficult) processing is required to integrate the information from multiple sensors (e.g., the *correspondence problem* in stereo and motion analysis).

Optimization approaches have been widely used to realize robust processing of signal data, for example by edge detection (Mont1971) and matching by dynamic programming (Fisc1973, Ohta1985), and optimization of a smoothness criterion in computing optical flow (Hild1983). In general, the performance of optimization methods is heavily dependent on the evaluation functions to be optimized. Although even simple evaluation functions work well at signal level processing, it is very difficult to formulate evaluation functions for interpretation; objects with complex structures and their mutual relations should be adequately represented in the evaluation functions. Moreover, optimization methods often perform uniform processing, while the image structure and complexity of the scene vary depending on locations. Thus it is difficult to realize adaptive analysis with optimization methods alone.

Using numerical values (probability) to represent reliability (or certainty) is very common in image understanding. Such probability computation is used at various stages in IUSs:

1. Information from multiple different sources (e.g., sensors) can be integrated by numerical computation.
2. During searching, we can select the most promising path to pursue based on the associated reliability (cost and utility).
3. We can select the most plausible interpretation by comparing numerical reliabilities associated with possible interpretations.

Reasoning with uncertain information is called *probabilistic reasoning*: probabilistic logic combines logical inference with probability computation (Gene1987); many expert systems use certainty factors to represent uncertain knowledge (Haye1983), and fuzzy sets characterize ambiguities involved in symbolic (qualitative) words (Zade1965).

Among various methods in probabilistic reasoning, *evidential reasoning* is useful to cope with incomplete information. In evidential reasoning, each piece of information is associated with a probability representing its reliability. When new information is obtained, the probability is modified depending on whether or not the new information is consistent with it. Thus the kernel computation in evidential reasoning is how to modify the probability to accommodate new information. In other words, evidential reasoning assumes that complete information is not always available, and computes reliabilities based on currently available partial information.

In statistical pattern recognition and other areas in computer science, the Bayesian probabilistic model has been widely used (Duda1973). However, it was pointed out that this model is not suitable for evidential reasoning (Lowr1982); it cannot discriminate *disbelief* and *lack of belief*. In the Bayesian model, when no information is available to decide whether or not proposition A is true, both $P(A)$ and $P(\neg A)$ are set to 0.5. This is because of the fundamental constraint $P(A) + P(\neg A) = 1.0$, where $P(A)$ denotes the probability of A and $\neg A$ denotes A's negation. However, what we want to represent is not believing A and $\neg A$ with equal probability but the lack of belief.

Recently, Dempster and Shafer theory was introduced to cope with this problem in evidential reasoning (Garv1981, Barn1981, Lowr1982). In this theory, the probability is represented by a subinterval [SUPT, PL] within the unit interval [0.0, 1.0], where $0.0 \leq \text{SUPT} \leq \text{PL} \leq 1.0$. SUPT and PL represent lower and upper probabilities, respectively, with which the proposition is believed: [1.0, 1.0] implies complete belief, [0.0, 0.0] complete disbelief, and [0.0, 1.0] complete lack of belief. Intuitively, at the initial stage the probability of each piece of information is set to [0.0, 1.0], and if a piece of evidence supporting the proposition is newly obtained, SUPT is increased. On the other hand, if new evidence is conflicting the proposition, PL is decreased. Each inserted piece of evidence has its own probability, and the integration between probabilities of the current and newly inserted evidence is performed by Dempster's theory of combination. Several methods have been proposed to incorporate this theory into IUSs (Faug1982, Wesl1982, Garv1987, Andr1988).

Although probabilistic reasoning is useful to cope with incomplete information, it is *passive*; no active reasoning mechanism to generate hypotheses for missing information is included. As discussed before, in image understanding we need active reasoning mechanisms to complement the intrinsic insufficiency of the input information. Thus in general probabilistic reasoning should be combined with such active reasoning.

An implement characteristic of knowledge-based analysis is that we can complement the intrinsic insufficiency of the input information by providing the information stored in the world model, that is, we can actively generate hypotheses based on the stored knowledge.

Moreover, knowledge-based analysis is also useful to complement the imperfection of segmentation. The Hough transform (Duda1973) and the generalized Hough transform (Ball1981) are good examples. They actively generate hypotheses (i.e., trajectories in the parameter space) based on object models (i.e., analytic equations representing object shapes). Then, mutually consistent hypotheses are accumulated (i.e., accumulation points in the parameter space are detected) to extract objects in the image. Thus we can detect complete straight lines, circles, and arbitrarily shaped objects even if some parts of them are missing and erroneous features are included in the image. These advantages come from the active generation of hypotheses based on the object models, a step which is not included in probabilistic reasoning or optimization methods. As is well known, however, the Hough transform requires both much computation time to generate hypotheses and large memory space to record all generated hypotheses. As will be described in Chapter 2, our spatial reasoning method used in SIGMA shares much with the Hough transform—both advantages and disadvantages.

From a philosophical point of view, knowledge-based analysis can be considered as a framework within which to tackle a profound problem in visual perception: the relation between *sensation* and *perception*. Sensation can be characterized by instantaneous absolute measurement, perception by persistent relative recognition. (Here we use *recognition* in a wide sense that includes the grouping process as well as the object recognition process.) The former changes depending on sensing devices, environments (e.g., lighting conditions), and time of measurement, and produces absolute quantitative signals, while *percepts* (i.e., what are perceived) are stable and formed relatively in the context defined by other percepts. In other words, we are always seeing objects we have never seen before at the sensation level, while we perceive familiar objects everywhere at the perception level. In his book (Greg1970) Gregory wrote

> perception is not a matter of sensory information giving perception and guiding behaviour directly, but rather the perceptual system is a "look up" system in which sensory information is used to build gradually, and to select from, an internal repertoire of "perceptual hypotheses"—which are the nearest we ever get to reality.

In knowledge-based analysis, object models are used to fill the gap between sensation and perception. In other words, these two processes interact through object models: sensation instantiates object models to generate (hypothetical) object instances, in terms of which percepts are constructed. Since the sensory information is used only to trigger the instantiation of object models, it need not contain complete information about the scene.

Note that since the input sensory information is limited, many competing object instances are usually generated and their number increases as the size of the world model grows. Thus we may need additional types of information to select plausible ones: objective, preference, prejudice, and belief.

Problems in knowledge-based analysis include what and how much information we should store in the world model and how efficiently we can use the sensory information to select and instantiate appropriate object models. It is one of our research goals to formulate a theory to solve these problems.

Although many schemes for knowledge representation and reasoning have been developed in AI (Nils1980), ordinary deductive reasoning based on classical formal logic (i.e., first-order predicate logic) does not work well in image understanding. This is because it assumes that all necessary information required for inference is given *a priori,* while in image understanding this assumption does not hold; *facts* about the world (i.e., observed information) are incomplete. In other words, ordinary logical reasoning proves the validity of theorems in the light of the given (fixed) information, while IUSs need to construct the description of the scene by dynamically generating new information. Thus we need to develop new schemes of knowledge representation and reasoning to derive meaningful conclusions with incomplete information. We can call such reasoning *constructive reasoning.* We will discuss this problem in more detail in Section 2.3.

1.4. SPATIAL REASONING IN IMAGE UNDERSTANDING

Since geometry is an important axis defining the world, spatial reasoning is required in various task domains, for example, image understanding, robot navigation, computer-aided design, and geographic information processing. In this section, we summarize various schemes of representing spatial relations and spatial reasoning methods for image understanding. (We use *spatial relations* in a general sense and *geometric relations* to denote specific spatial relations involving metric information.)

Topological relations, such as adjacency, intersection, and inclusion, have often been used in IUSs to represent spatial relations among image features and objects (Ohta1980, Rubi1980). They also play an important role in describing the structures of objects [e.g., the winged edge model for 3D objects (Ball1982)]. Although topological relations are useful, they are not enough to characterize structures of and spatial relations among geometric entities. Most of them are defined between *connected* geometric entities, yet there are many geometric relations, involving distance and directional information, which are defined between disjoint entities: left/right, above/below, proximate, parallel, colinear, coplanar, and so on.

In image processing, several data structures are used to characterize geometric relations among *disjoint* image features: the minimum spanning tree (Zahn1974), the Voronoi diagram, the Delaunay triangulation (Sham1975), the quad tree, and the *k-d* tree (Bent1979). These data structures were originally devised to describe structurally spatial proximity among a set of points, and later generalized for lines and regions. Zahn (Zahn1974) represented the spatial proximity among data points by their minimum spanning tree and used its structural features (e.g., angles between branches) for matching. Note that in his method geometric relations themselves are used as matching keys rather than attributes of image features (Mats1984c). Toriwaki et al. (Tori1982) and Matsuyama et al. (Mats1984a) developed algorithms for computing digital Voronoi diagrams for disjoint lines and regions, and characterized several types of proximity relations among these image features. Matsuyama et al. (Mats1984b) applied the *k-d* tree to implement an efficient file system for map database systems. Proximity relations defined by these data structures complement ordinary topological relations. Lowe (Lowe1987) used colinearity, proximity, and parallelism to group line segments.

In early IUSs, geometric relations were often represented by defining symbolic descriptive terms for them. For example, Winston (Wins1975) defined a 2D relation *LEFT-OF* as follows:

A is *LEFT-OF* B
if the centroid of A is located at the *left of* that of B, and
if the rightmost point in A is located at the *left of* that in B.

left of in the above conditions is defined simply by comparing X (horizontal) coordinate values. Although the geometric meanings of various descriptive terms, such as *ABOVE* and *BETWEEN*, were investigated (Free1975), it is difficult to define them consistently; *LEFT-OF* must be the reverse relation of *RIGHT-OF* [i.e., *LEFT-OF*(A, B) must be equivalent to *RIGHT-OF*(B, A)], and only one of *LEFT-OF*, *RIGHT-OF*, *ABOVE*, and *BELOW* should be satisfied between a pair of geometric entities.

The problem of defining such descriptive terms involves a difficult universal problem of interfacing numerical (quantitative) computation with symbolic (qualitative) computation. We often encounter the same problem in developing expert systems (Haye1983), in which knowledge is represented by a set of rules. Since rules are to be described in terms of predefined symbolic vocabulary, what descriptive terms we should prepare to characterize various quantitative features in the task domain and how to define their meanings are crucial problems in developing capable expert systems. How to describe and reason about quantitative characteristics of various physical systems (e.g., electric circuits) based on qualitative symbolic features is a primary research objective of *qualitative resoning* in AI (Spec1984).

One idea of integrating quantitative and qualitative information is to incorporate *fuzzy* descriptions (Zade1965). Haar (Haar1982) used fuzzy predicates to characterize the ambiguity involved in descriptive terms for spatial relations. A spatial relation is described by using two primitive fuzzy predicates, DISTANCE and BEARING:

$$\{(DISTANCE \ A \ B \ (7 \ 10)) \ 0.6\}$$
$$\{(BEARING \ A \ B \ (45 \ 60)) \ 0.8\}$$

The former means that the distance between A and B is between 7 and 10 and the latter that the direction from A to B is between 45 and 60. Each proposition is associated with its reliability: 0.6 and 0.8 in the above examples. Quaitative terms like LEFT-OF for spatial relations are defined using these fuzzy descriptions. In Haar's system several computational rules are implemented to perform reasoning about object locations based on fuzzy descriptions.

In spatial reasoning in 2D space, *iconic* descriptions have been often used to specify object locations. In Haar's system, upright rectangular regions were used to describe estimated (approximate) object locations. McDermott (McDe1980b) used a similar method to reason about locations of geographic objects in maps. Russell (Russ1979, Ball1982) used a *constraint (location) network* to represent spatial relations among objects in aerial photographs. The network in Fig. 1.5 represents two spatial relations associated with aeration tanks:

1. *Aeration tanks are located somewhere close to both the sludge tanks and the sedimentation tanks.*
2. *Aeration tanks must not be too close to either the sludge tanks or the sedimentation tanks.*

The system evaluates the network to reason about the locations of aeration tanks: object locations are represented by 2D regions and the system performs set operations among them. Given location data (i.e., region) for a certain data node in the network (e.g., sludge tank), each computation node (e.g., union) performs its corresponding set operation to produce new data, which then are used as input to another computation node.

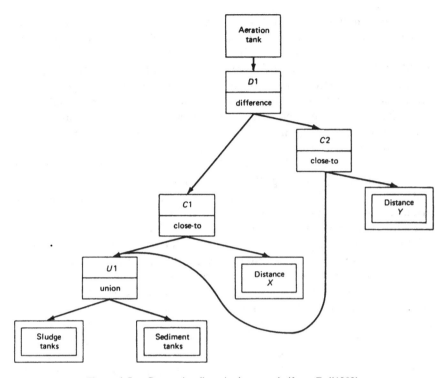

Figure 1.5.   Constraint (location) network (from Ball1982).

The most general method of representing geometric relations would be to use geometric transformations between object-centered coordinate systems defining structures of individual objects. In ACRONYM (Broo1981), a geometric relation between parts of objects, which are described by generalized cylinders (Fig. 1.4), is represented by a transformation matrix which transforms one object-centered coordinate system into another. One representational problem in this method is that, in addition to the locational relation defined by the transformation matrix, the shape information of the connecting portion (joint) between the parts should be included in the object model. Since joints are often projected onto images as very informative image features (i.e., corners), models of objects should include their explicit descriptions to facilitate matching with image features. Marr (Marr1978) discussed types of joints between a pair of generalized cylinders. Fisher's 3D object modeling system (Fish1987) allows descriptions of joints between parts of objects.

ACRONYM represents the ambiguity involved in spatial relations by inequalities (constraints) and performs spatial reasoning using a *constraint manipulation system* (CMS). When an image feature is matched with a part of an object, attributes of the image feature are assigned to corresponding variables in the abstract object model (e.g., FUSELAGE-RADIUS and FUSELAGE-LENGTH in Fig. 1.4a), and an instance of that object model is created. However, since image features are 2D while object models are 3D, we cannot determine exact values of all variables in the model. Thus, the object instance itself is associated with a set of constraints (inequalities) including variables. CMS examines the consistency between each pair of object instances to construct a scene description. First it applies to one of the object instances the geometric transformation representing the spatial relation between them. Then CMS merges the set of constraints associated with the transformed instance and the constraint set associated with the other instance. It examines the question of whether the merged constraint set can have solutions; since constraint sets are usually underconstrained, CMS only examines the existence of solutions. If the merged constraint set can have solutions, the pair of object instances are regarded as consistent. ACRONYM integrates consistent object instances to construct a scene description. The more consistent instances are merged, the tighter the set of merged constraints becomes.

As discussed in Section 1.2, one serious problem in ACRONYM is that its reasoning is purely bottom-up and no top-down analysis for detecting missing objects is incorporated. In other words, ACRONYM uses spatial relations only to examine the consistency between object instances, whereas the same relations can be used to generate expectations about missing objects in the top-down analysis. In the next section, we will provide an overview of bottom-up and top-down analyses in IUSs, and we will discuss the use of spatial relations in these two analysis processes in Section 2.2.

1.5. CONTROL STRUCTURES FOR IMAGE UNDERSTANDING SYSTEMS

How to control reasoning processes is an important problem in realizing intelligent computer systems. In AI, many search algorithms have been proposed to realize efficient reasoning (Nils1980). Sometimes *control knowledge* (i.e., knowledge about how to use the knowledge about the world) is required to realize flexible control structures. Reddy (Redd1979) pointed out that "Knowledge reduces uncertainty and, therefore, search, and conversely, search can compensate for a lack of knowledge." In image understanding, since we have to perform reasoning faced with a lack of sufficient information, searching among a huge number of alternatives is inevitable, and therefore the problem of controlling the search is crucial.

In general, IUSs can construct the scene description by using one of the following two types of analysis:

1. *Bottom-up analysis* (*data-driven analysis*). The input information is gradually organized from raw image data to image features and then to the structural description of the image. Then the matching process establishes the correspondence between image features and object models to generate the description of the scene (Fig. 1.2). For example, to first extract a rectangle from an aerial photograph and then recognize it as an instance of a house is a bottom-up analysis.

2. *Top-down analysis* (*model-driven analysis*). First the appearance of the object to be recognized is determined and that information is used to extract image features (Fig. 1.2). In recognizing a house, for example, first IUSs instantiate the house model and activate the image analysis module to detect a rectangle corresponding to the instantiated house model.

Some IUSs have been designed based on bottom-up analysis, in which relaxation labeling (constraint filtering/satisfaction) (Walt1975, Rose1976, Humm1983) was often used as the reasoning mechanism. In MSYS developed by Tenenbaum and Barrow (Tene1977), the image is first segmented into regions and a set of initial labels representing all possible interpretations is assigned to each region. Then, geometric constraints between labels (i.e., spatial relations between objects) are used to filter out inconsistent labels. ACRONYM also uses this filtering mechansim to construct scene descriptions.

Although relaxation labeling (in a general sense) is a useful computational scheme, we encounter serious problems when we try to use it as the control structure for object recognition:

1. *Poor descriptive capability*. The descriptive terms in relaxation labeling are limited to labels and compatibility coefficients. They are too simple to describe knowledge about structures of objects and their mutual relations. Levels of abstraction and structural composition should be explicitly represented in the world model to attain versatile recognition capabilities.

2. *Uniform processing over the entire scene*. Uniform iterative probability modification (constraint propagation) is too simple to

cope with varieties of different local situations in a scene. In a
complex scene, constraints often change depending on local
contexts, so that focus of attention and context-sensitive local
analysis are required (Naga1984).

3. *Pure interpretive recognition.* Recognition by relaxation labeling
   is not constructive but interpretive; as discussed before, image
   understanding is not the simple process of labeling image features
   but the process of constructing the scene description. Especially
   since the input information obtained from the image is limited,
   IUSs should actively generate missing information based on the
   world model. Relaxation labeling involves no such constructive
   reasoning process.

Note that we are not saying that relaxation labeling is useless but rather
that it is too simple to serve as the control structure for IUSs. Havens and
Mackworth (Have1983) also pointed out limitations of relaxation
labeling.

Several workers have tried to augment the descriptive power of
relaxation labeling. Davis and Rosenfeld (Davi1978) and Price (Pric1982)
introduced composition hierarchies based on *PART-OF* relations to realize
hierarchical (multilevel) relaxation. Tsotsos (Tsot1981) proposed an
iterative procedure to compute reliabilities of objects. Objects in his
world model are organized into a hierarchical network based on
A-KIND-OF, *PART-OF*, and *SIMILARITY* relations. Compatibilities between
objects are computed based on the conceptual adjacency defined by the
relations. He demonstrated that bottom-up and top-down propagations
of reliability values through the relational links facilitate fast
convergence.

Some IUSs have been designed based on top-down analysis. Bolles
(Boll1976) proposed a *verification vision system* for industrial part
recognition. It first constructs a goal, specifying the target object to be
recognized. The goal is represented by a template and is matched with
the image. Selfridge (Self1982) emphasized the utility of top-down
analysis in his aerial image understanding system, into which *failure-
driven* reasoning is incorporated to select appropriate object appearances
and image analysis procedures are used in the top-down analysis. The
system first uses a plausible appearance and a possible image analysis
procedure to detect the target object. When the first top-down analysis
fails, it reasons about causes of the failure and modifies the appearance
and/or the procedure. Rosenthal (Rose1978) used two hierarchies to
realize coarse-to-fine top-down analysis: a conceptual hierarchy describing conceptual relations among object models and a resolution hierarchy

consisting of images with different resolutions. When a query about the target object is given, the system starts the top-down analysis at the top of the hierarchies. By descending the hierarchies, the concrete object model and its exact location are gradually determined. VISIONS (Hans1978) uses similar hierarchies to conduct coarse-to-fine reasoning. Russell (Russ1979) used a constraint network (Fig. 1.5) to reason about the locations of objects in the top-down analysis. Top-down analysis enabled these systems to realize reliable image segmentation.

In general, two types of reasoning are required to realize effective top-down analysis:

1. *Model selection.* Since a variety of object models are included in the world model, we have to select appropriate object models to guide the top-down analysis.
2. *Spatial reasoning about expected object locations.* In order to realize efficient and effective top-down analysis, we have to determine the approximate locations of target objects; since structures of local areas in the image are simple, the top-down image analysis can use simple image models to extract image features.

We need to incorporate capable *focus of attention* mechanisms to make the top-down analysis—determination of appropriate object models and estimation of their location—work effectively. Rich *source* information is required for this focusing process and, therefore, the problem of the top-down analysis is how IUSs can obtain such information.

Currently, it is generally accepted that IUSs should incorporate both bottom-up and top-down analyses, since they complement each other:

1. The bottom-up analysis is general but is not flexible enough to achieve recognition which adapts to the structure of the scene. In particular, it is very difficult for the uniform image processing in the bottom-up analysis to extract meaningful image features perfectly.
2. Although adaptive context-sensitive recognition and reliable image segmentation can be realized by the top-down analysis, it requires rich source information to focus its attention.

Thus it is natural to design IUSs which first apply the bottom-up analysis and then activate the top-down analysis based on the partial information obtained by the bottom-up analysis. VISIONS by Hanson and Riseman (Hans1978), Ohta's system (Ohta1980), and our former aerial image understanding system (Naga1980) used both types of analyses. However, they used ad hoc rules to determine which type of analysis was to be used at what stage during the entire analysis process. Moreover, the capability of the top-down image segmentation for missing objects was limited. Thus, a major problem in designing IUSs is how to integrate these two analysis processes. Note that although the use of rich control knowledge is useful to realize flexible control structures, they usually represent ad hoc knowledge specific to application domains. What we really want here is a general reasoning framework for IUSs.

In this book, we propose a general reasoning framework which enables IUSs to integrate both bottom-up and top-down analyses. We have developed an image understanding system, SIGMA, based on this framework and demonstrate its performance in analyzing images of suburban housing developments.

1.6. CONTENT OF THE BOOK

In the rest of the book, we describe our image understanding system, and we propose the following methods to cope with the three problems in IUSs discussed in Section 1.3:

1. *Evidence accumulation for spatial reasoning.* In SIGMA reasoning for object recognition is mainly performed based on spatial relations between objects, while our former system recognized objects using their spectral and shape features. When an instance of an object (e.g., a house) is recognized, the system generates hypotheses about its related objects (e.g., neighboring houses and a facing road). That is, in SIGMA spatial relations are used actively to generate hypotheses. We call object instances and hypotheses *evidence.* Pieces of evidence from different sources are accumulated to establish relations between object instances or to search for missing objects. The search process activates the segmentation expert [see (3) below] to extract new image features corresponding to the missing objects. The accumulation of evidence decreases the total amount of effort spent in the search and increases the reliability of the analysis. In short, the reasoning process actively generates new information (i.e., hypotheses) based on object models, which complements the insufficiency of the information obtained from the image. Moreover, our method of evidence accumulation enables us to integrate both bottom-up and top-down analyses into a unified reasoning process.

2. *Distributed problem solving based on object-oriented knowledge representation.* In SIGMA, all recognized objects perform reasoning

about their surrounding environments based on their own knowledge. This means that each object instance is regarded as an independent reasoning agent. In order to realize such *distributed problem solving* capability, we used knowledge representation based on *object-oriented computation* (Gold1983). The knowledge about a class of objects in the world is stored in an *object class* along with a set of production rules. When an object in the scene is recognized, the corresponding object class is instantiated to generate an object instance. Each object instance uses the set of production rules stored in the object class to generate hypotheses about its related objects and to perform reasoning based on whether or not the hypotheses are verified. The system coordinates such local reasoning by independent reasoning agents (i.e., object instances) to construct a globally consistent description of the scene.

3. *Expert system for top-down image segmentation.* We can realize reliable segmentation if the quality (structure) of image data is uniform (simple) and/or models of image features to be extracted are given. In SIGMA, top-down image segmentation is extensively used to extract image features which were not detected by the initial segmentation. The object recognition module first determines local areas for the top-down segmentation and specifies models of image features to be extracted to the segmentation module. Since the structures of local image data are simple and models of target image features are given, efficient and reliable segmentation can be realized.

The segmentation module in SIGMA is an *expert system for image processing* which reasons about the most effective segmentation methods by using knowledge about image processing techniques. It first reasons about the most effective analysis plan. Then, it selects appropriate image processing operators, determines optimal parameters based on the image quality, and modifies operators and adjusts parameters in case of failure. This expert thus realizes flexible automatic image segmentation.

In Chapter 2, we first describe the overall architecture of SIGMA. SIGMA consists of four experts (i.e., reasoning modules): the Geometric Reasoning Expert for spatial reasoning, the Model Selection Expert for appearance model selection, the Low-Level Vision Expert for knowledge-based segmentation, and the Question and Answer Module for interactive information retrieval. Sections 2.1.2–2.1.4 describe the tasks of and mutual relations between these experts. Sections 2.2 and 2.3 are devoted to the process of evidence accumulation for spatial reasoning in SIGMA. We first describe the motivations and principles used in evidence accumulation with illustrative examples. SIGMA uses spatial relations between different types of objects and *PART-OF* relations to describe composite objects consisting of many constituent parts. In

Section 2.2.3, we present detailed reasoning processes based on PART-OF relations. Then, in Section 2.3, we discuss the logical foundations of our spatial reasoning in terms of the first-order predicate calculus. Section 2.4 describes the knowledge representation scheme used in SIGMA for representing the world model. The object-oriented knowledge representation is described with illustrative examples.

Chapter 3 describes practical algorithms used to realize the reasoning mechanism. Section 3.1 provides an overview of the reasoning processes performed by the Geometric Reasoning Expert. Sections 3.2 and 3.3 describe the process of hypothesis generation and the representation of evidence (i.e., object instances and hypotheses), respectively. Sections 3.4–3.6 present three kernel processes of our spatial reasoning: consistency examination among pieces of evidence, the focus of attention mechanism to control the interpretation process, and reasoning processes to establish relations and to activate the top-down analysis. In Sections 3.7 and 3.8, many illustrative examples of production rules for various types of reasoning are demonstrated. A detailed example of the entire spatial reasoning process is given in Section 3.9.

In Chapter 4, we describe the knowledge representation and reasoning mechanism used in the Low-Level Vision Expert. Section 4.1 gives a survey of expert systems for image processing which analyze images guides by knowledge about image processing techniques. Sections 4.2 and 4.3 respectively describe the knowledge representation and the reasoning process of our segmentation expert. In Section 4.4, several experimental results of the top-down image segmentation are given to demonstrate the capability of the expert. We emphasize the importance of *image analysis strategies* to realize effective image analysis and propose two schemes of representing such strategies in Section 4.5. In Section 4.6 several future problems for expert systems for image processing are discussed.

In Chapter 5, the experimental results of aerial image understanding by SIGMA are given. Several intermediate analysis results are shown to illustrate the reasoning processes of SIGMA in various situations. A simple question and answer module was developed to examine the content of the scene description constructed by SIGMA. This module was used to generate all results shown in this chapter. In Section 5.5, results of the analyses of several different aerial photographs are given to demonstrate the performance of SIGMA.

In Chapter 6, we summarize the ideas and algorithms used in SIGMA and discuss future problems.

Chapter 2

System Architecture and Reasoning Scheme in SIGMA

In this chapter, we first describe the architecture of SIGMA. It consists of the following four reasoning modules: the *Geometric Reasoning Expert* for spatial reasoning, the *Model Selection Expert* for appearance model selection, the *Low-Level Vision Expert* for knowledge-based image segmentation, and the *Question and Answer Module* for examining the content of the constructed description of the scene. We summarize their roles, knowledge sources, and reasoning methods. Then, we present the principle behind our spatial reasoning, which we call *evidence accumulation,* and describe how the experts listed above cooperate to construct the scene description. Following several illustrative examples, we discuss the logical foundations of our reasoning method in terms of the first-order predicate calculus. In the last section, we present the knowledge representation scheme for describing the world model: the structures of object models, their relations, and the control knowledge.

2.1. ORGANIZATION OF THE SYSTEM

2.1.1. Design Principles

Figure 2.1 illustrates the overall configuration of SIGMA. It consists of four reasoning modules: the *Geometric Reasoning Expert* (GRE), the

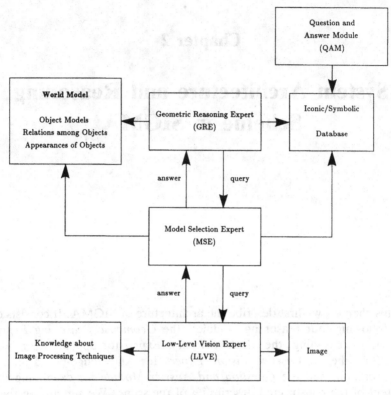

Figure 2.1. Organization of the system.

Model Selection Expert (MSE), the *Low-Level Vision Expert* (LLVE), and the *Question and Answer Module* (QAM). The former three modules are analysis modules to construct the description of the scene, and the last realizes the interactive information retrieval facility to examine the content of the constructed description. We use the term *experts* to denote the three analysis modules. Since our primary concern is how to construct the description, most of the discussions in this book deal with these experts, while we extensively used QAM to obtain the experimental results illustrated in Chapter 5.

As discussed in Section 1.1, image understanding systems (IUSs) need three different types of knowledge: scene domain knowledge, image domain knowledge, and knowledge about the mapping between the scene and the image. We incorporated three independent experts corresponding to these three types of knowledge; each expert performs reasoning using its corresponding type of knowledge and communicates with the others via simple question and answer protocols. That is, GRE uses scene domain knowledge for spatial reasoning among objects, MSE reasons about appearances of objects in the image based on knowledge about the mapping, and LLVE performs image segmentation using image domain knowledge.

As illustrated in Fig. 2.1, the three experts of SIGMA are loosely coupled. This system organization is quite different from that of our former image understanding system based on the blackboard model (see Fig. 1.3). This difference stems from the design principles to which we adhered in creating SIGMA:

1. Since the problem of image understanding is very complicated, it is difficult to design a totally integrated system architecture from scratch. Instead, we should divide the problem into independent subproblems and should investigate what functions are required to solve each subproblem. Since the knowledge required for image understanding can be classified into three types and the type of knowledge defines the task and style of the reasoning to be performed based on that knowledge, we incorporated an independent reasoning module (i.e., expert) corresponding to each type of knowledge. Thus each expert performs reasoning based on its own type of knowledge. It is our claim that three different types of reasoning are required in image understanding: spatial reasoning, appearance model selection, and image analysis. (Readers may wonder why image analysis is a reasoning process. We will discuss what reasoning is to be performed in image analysis in Sections 2.1.4 and 4.1.)

2. As will be described below, SIGMA extensively uses the top-down analysis to construct the description of the scene. MSE and LLVE perform two subtasks required in the top-down analysis: appearance model selection and image analysis to detect missing objects. That is, GRE is the major reasoning module in SIGMA and the other two experts can be considered as its submodules.

Although the computation speed may be decreased by this loosely coupled architecture, it has the following advantages:

1. We can discriminate clearly the types of knowledge, which were often confused in early IUSs. In our former system, for example, the reasoning based on image domain knowledge was encoded in image analysis programs, and that based on knowledge about the mapping was not used: all reasoning and processing were done at the image level.

2. Since the task and the knowledge source for each expert are precisely defined, we can devote ourselves to investigate what reasoning and computation are to be performed to fulfill the required task. That is, our research goal is not to develop practical IUSs but to study what types of reasoning are essential in IUSs.

3. Since each expert can be designed independently of the others, we can use the most suitable architecture for it. That is, a given expert can be designed based on the blackboard model, and others based on production systems, semantic networks, and frames (see Ball1982, Haye1983, and Mins1975 for discussions of these knowledge representation schemes).

4. To the extent that the interfaces between the experts are consistent, we can modify each expert without affecting the others. This is very convenient to enable us to develop the system incrementally.

In what follows, we provide an overview of the task and reasoning method of each expert.

2.1.2. Geometric Reasoning Expert

GRE is the central reasoning module in the system: it constructs the description of the scene by establishing spatial relations among objects. It uses as its knowledge source a symbolic hierarchical world model representing geometric structures of and spatial relations among objects. We call the spatial reasoning done by this expert *evidence accumulation*, in which both bottom-up and top-down analyses are integrated into a unified reasoning process.

As will be described in detail in Section 2.4, we use an *object-oriented paradigm* (Wein1980, Gold1983) for representing the world model. Object-oriented knowledge representation is very similar to that based on *frames* (Mins1975). In SIGMA, the knowledge about a *class* of objects in the world is represented by an *object class*, in which the attributes of and relational information about that class of objects are stored. A recognized object is represented by an *object instance*: when an object in the scene is recognized, the object class is *instantiated* to produce the object instance representing the recognized object. (When it is clear from the context, we sometimes use *objects* to refer to both object classes and object instances. *Object models* and words in small italic capital letters like *HOUSE* denote object classes.)

In SIGMA, each object instance itself is an active reasoning agent which generates *hypotheses* for its related objects based on the knowledge stored in the corresponding object class. For example, a house instance generates hypotheses for its neighboring houses and the facing road. It is this reasoning capability that discriminates object-oriented knowledge representation from that based on frames: objects in the former are active reasoning agents, while frames are static data structures manipulated by a reasoning engine. Thus, the reasoning for constructing the scene description is conducted by both individual object instances and GRE: each object instance performs reasoning about its local environment, while GRE coordinates such local reasoning activities to construct the globally consistent description of the scene. In this sense, we can consider our spatial reasoning as *distributed problem solving* (Smit1978), in which many independent reasoning agents (i.e., object instances) cooperative to solve a large problem (to construct the scene description).

We call object instances and hypotheses generated by them *evidence*. All partial evidence obtained during the interpretation process is stored in *the Iconic/Symbolic Database* (Fig. 2.1), where consistent pieces of evidence are accumulated. Each piece of evidence is described in two different forms: the ionic description for representing its location (i.e., occupied space) and the symbolic description for representing its attributes and relations to others.

GRE first examines the consistency among all pieces of evidence in the database to form what we call *situations*. Each situation consists of mutually consistent pieces of evidence and represents a local environment (context). GRE selects one situation and focuses its attention on the local environment represented by the selected situation. Then, either the bottom-up analysis to establish a relation between object instances or the top-down analysis to search for a new (missing) object is activated, depending on the nature of the local environment on which it is focusing. (Detailed descriptions of our evidence accumulation method will be given in Sections 2.2 and 2.3.)

In the top-down analysis, GRE first reasons about its goal: the target object to be detected and where in the scene to analyze. Then it asks MSE to perform the object detection according to the goal specification. For example, Fig. 2.2a shows a goal specification given to MSE. It reads as follows:

1. The class of the target object is *HOUSE* and its area is between 250 m^2 and 500 m^2.
2. It is located in the rectangular area whose upper left and lower right corners are $(100, 300)$ and $(600, 800)$ respectively. [Here we consider the scene to be two-dimensional (2D).]
3. The context in the specification denotes a list of object instances suggesting the existence of the target object.

```
a ((GOAL        (AND (EQUAL OBJECT-TYPE HOUSE)
                     (AND (LESSP AREA 500)
                          (GREATERP AREA 250))))
   (LOCATION (AND (LESSP X 600)
                  (GREATERP X 100)
                  (LESSP Y 800)
                  (GREATERP Y 300)))
   (CONTEXT  (HOUSE_GROUP002 HOUSE_GROUP006 ROAD005)))


b ((GOAL        (AND (EQUAL IMAGE-FEATURE-TYPE RECTANGLE)
                     (AND (LESSP AREA 200)
                          (GREATERP AREA 100))))
   (LOCATION (AND (EQUAL START-I 40)
                  (EQUAL START-J 75)
                  (EQUAL END-I  240)
                  (EQUAL END-J  320))))
```

Figure 2.2. Goal specifications in the top-down analysis. (a) Goal specification to MSE. (b) Goal specification to LLVE.

Note that. all properties and the coordinate system used in the goal specification are described in terms of the terminology defined in the scene domain.

If the target object is successfully detected by MSE, GRE establishes spatial relations between that new object instance and those in the context (i.e., already recognized object instances which suggested the existence of the detected object instance). Otherwise, GRE reports the failure to all object instances in the context, which then would modify their hypotheses. This failure-driven reasoning is conducted by each object instance based on the knowledge stored in its corresponding object class.

2.1.3. Model Selection Expert

MSE is activated by GRE to reason about the most promising appearance of the target object to be detected in the top-down analysis. After determining the appearance, MSE activates LLVE to extract one or more image feature(s) which matches the selected appearance. If all these processes are completed successfully, MSE generates an object instance representing the target object, and returns it to GRE as the answer to the request. The appearance determination involves the following three processes:

1. *Selection of a specialized object model.* In general, GRE performs reasoning based on spatial relations defined among *generic* objects, in terms of which the goal specification of the top-down analysis is described. For instance, "a *HOUSE* is facing a *ROAD*" and "a *DRIVEWAY*

connects a *HOUSE* with its facing *ROAD*" are examples of spatial relations, and "find a *HOUSE*" is an example of a top-down request. However, these generic objects (i.e., *HOUSE*, *ROAD*, and *DRIVEWAY*) usually refer to abstract objects with no specific geometric structures (i.e., shapes). For example, generic object *HOUSE* denotes the generalized concept of many types of houses with specific shapes. Thus, in order to recognize the target object, MSE has to reason about a specialized object model with a concrete shape.

In SIGMA, object models in the world model are organized on several levels of specification: a generic object model at the top level is gradually specialized to specific object models. We use *A-KIND-OF* relations to represent generalization/specialization relations among object models. (See Section 2.4 for a discussion of the knowledge organization in SIGMA.) In general, most specialized object models include concrete geometric structures. For example, *HOUSE* is specialized to *RECTANGULAR-HOUSE* and *L-SHAPED-HOUSE*, each of which contains concrete shape information: rectangle or L-shape.

Since GRE specifies the goal in terms of generic objects (Fig. 2.2a), MSE first reasons about which specialized object model is to be searched for in the image. "Find a *HOUSE*" is too general to guide the top-down image analysis, because there are many house models with different shapes. MSE uses the contextual information in the goal specification for this reasoning: since the context describes the local environment in which the target object is embedded, we can reason about the specialized object model most plausible in the light of its surrounding environment. For example, suppose the target object is a *HOUSE* and the context consists of its neighboring houses. The neighboring houses refer to already recognized object instances with concrete shapes. It would be reasonable to consider the properties (e.g., shape) of the target house to be similar to those of its neighbors: if most of the neighbors are rectangular, assume the target house to be *RECTANGULAR-HOUSE*.

As will be described in Section 2.4, each object class representing a generic object model contains knowledge about how to select a specialized object model. We call such knowledge *specialization strategies*, based on which MSE selects the most plausible specialized object model.

Generally speaking, the richer the context is, the more accurately can a model be determined in this model selection. However, since the knowledge used for this reasoning is heuristics with no concrete theoretical basis, the reasoning may sometimes fail: the specialized model determined by MSE may be found to be wrong and other models must be tried. This implies that, in general, MSE needs to perform a trial-and-error search to determine the appropriate specialized object model.

2. *Decomposition of a composite object model.* The target object requested by GRE may be a *composite object* consisting of many parts. In this case, MSE has to decompose the target object into its part objects; in the world model, the composite object is represented structurally in terms of its part objects and has no concrete shape by itself (see Fig. 1.4 for an example). For instance, when "find a CAR" is given as a goal specification, MSE decomposes CAR into TIRES, BODY, DOORS and so on, and asks LLVE to extract image features which match these part objects.

In SIGMA, a composite object is represented by an object class which is linked by PART-OF relations with a group of object classes representing its part objects. Each object class representing a composite object contains the knowledge about how to decompose it into its part objects. For example, "since a TIRE has a simple appearance (i.e., circle or ellipse), first ask LLVE to extract it from the image" is an example of such knowledge. We call such knowledge *decomposition strategies,* based on which MSE performs the reasoning about the appearance model selection for composite objects.

3. *Determination of 2D appearance.* An object model with concrete shape information is determined by the two reasoning processes described above. Next, MSE reasons about its 2D appearance: it transforms the object model in the scene domain into the appearance in the image domain.

Since all information in the goal specification given by GRE is described in terms of the terminology in the scene domain, MSE transforms it into that described in terms of the terminology in the image domain. To perform this transformation, MSE uses the camera model, which defines the geometric transformation between the scene and the image.

While in SIGMA, we assume the exact camera model is given a priori. It may be required to determine the camera model in some applications. In such cases, since the top-down analysis involves matching between expected models and image features, we can compute the camera model based on the matching results. Thus, if determination of the camera model is required, we must include that process in MSE.

After the reasoning processes described above, MSE generates a goal specification to LLVE. It includes the type of image feature to be extracted from the image and its properties (i.e., constraints to be satisfied). Figure 2.2b shows an example of the goal specification to LLVE. Note that all numerical values in this goal specification refer to properties and the coordinate system in the image domain.

If the image feature(s) representing the appearance of the target object is successfully extracted by LLVE, MSE generates an object instance representing the target object and returns it to GRE. Otherwise, MSE selects an alternative appearance model for the target object and repeats the same process. In general, this trial-and-error analysis should be controlled by the degree of the confidence with which GRE believes in the existence of the target object.

In summary, the task of MSE involves several different types of reasoning to determine the practical appearance of the target object. It also controls LLVE to extract image features which match the object appearance. Hierarchical knowledge organizations based on *PART-OF* and *A-KIND-OF* relations and knowledge about how to use these relations are required for MSE to perform the reasoning. Thus, the reasoning capability of MSE is heavily dependent on the knowledge included in the world model.

2.1.4. Low-Level Vision Expert

LLVE performs image segmentation to extract the image feature specified by MSE. Since there are many possible segmentation methods to extract the image feature from the image, LLVE reasons about which method is the most effective based on knowledge about image processing techniques.

Given a goal, LLVE first makes a global plan for the segmentation and selects appropriate practical image processing operators and determines their optimal parameters. For example, suppose the goal in Fig. 2.2b is given to LLVE. It first extracts the local area from the image specified in the goal, and reasons about a plan, such as Edge Detection → Edge Linking → Closed Boundary Detection → Polygon Approximation → Rectangle Recognition. Then a practical operator and its parameters are selected for each analysis step in the plan: Sobel operator for Edge Detection, eight-connectedness for Edge Linking, and so on. LLVE applies the selected operators to the local area to extract one or more rectangle(s) satisfying the constraints specified in the goal: in Fig. 2.2b, for example, the area of the rectangle must be between 100 and 200 pixels.

Since LLVE performs image analysis only in the local area specified by MSE, the selection of operators and parameters can be done rather easily; the image structure in the local area is very simple compared with that of the entire image. Moreover, LLVE performs trial-and-error analysis to evaluate the image structure (quality) and to adjust parameter values. All reasoning processes for plan generation, operator and parameter selection, and trial-and-error analysis are guided by knowledge about image processing.

Note that this knowledge is domain-independent and therefore LLVE can be used in IUSs for any type of scenes, while GRE and MSE use domain-dependent knowledge about specific applications. The incorporation of LLVE, an *expert system for image processing,* is one of the most important characteristics of SIGMA. It enables GRE and MSE to concentrate on their essential reasoning (spatial reasoning and model selection, respectively) without worrying about image segmentation: (Section 4.1 describes the background, objectives, knowledge representation, and reasoning methods used in expert systems for image processing in a general context.)

2.2. EVIDENCE ACCUMULATION FOR SPATIAL REASONING

In this section, we first discuss the use of spatial relations in the bottom-up and top-down analyses. Then we describe the spatial reasoning performed by GRE with illustrative examples. We call this reasoning *evidence accumulation for spatial reasoning*, in which both bottom-up and top-down analyses are integrated into a unified reasoning process. After describing spatial reasoning based on *PART-OF* relations, we discuss the logical foundations of our reasoning method in terms of the first-order predicate calculus.

2.2.1. Motivation

In general, two different types of information can be used to recognize an object: intrinsic properties (i.e., attributes such as size, shape, and color) and spatial relations to other objects. Our primary interest is in the representation of spatial relations and their utilization for image understanding. Here we call object recognition based on spatial relations *spatial reasoning*. This is especially important in the recognition of man-made objects; they are arranged according to well-organized geometric configurations. Moreover, although shape is often regarded as an intrinsic object property, a complexly shaped object is usually described structurally using spatial relations among its components. Thus structural shape analysis also requires spatial reasoning.

Consider a binary (spatial) relation $REL(O_1, O_2)$ between two object classes, O_1 and O_2. This relation can be used as a *constraint* to recognize

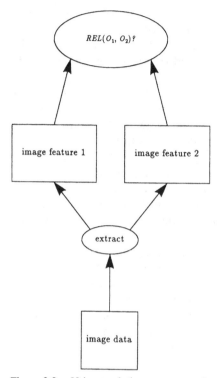

Figure 2.3. Using a relation as a constraint.

instances of these two classes (Fig. 2.3):

a. First extract image features which satisfy the intrinsic properties of O_1 and O_2. Then, we have two sets of image features corresponding to the object classes $\{a_1, a_2, \ldots, a_m\}$ and $\{b_1, b_2, \ldots, b_n\}$.
b. Next select a pair of image features (a_i, b_j) which satisfy the relation REL. (The mapping between the scene and the image is neglected.)

In this *bottom-up* recognition scheme, relations are used to *filter out* erroneous image features: image features which do not satisfy the relation (i.e., constraint) are considered as errors. This scheme assumes that a complete set of *correct* image features is extracted by the segmentation; the reasoning based on a relation cannot be performed unless a pair of correct image features corresponding to object instances are extracted. We can call this reasoning scheme *interpretive reasoning,* in which the complete set of correct image features to be interpreted is available and relations are used to sift out erroneous image features. This is the reasoning performed by relaxation labeling and constraint filtering (see Section 1.5).

The most critical problem in interpretive reasoning is that it is almost impossible to extract all meaningful image features by the initial segmentation; some correct image features are not extracted due to noise and the fact that parts of objects in the scene cannot be observed due to occlusions. Moreover, since in complex scenes the image quality varies depending on location and the information about such local image quality is not available during the initial segmentation, it is inevitable that some meaningful image features will be missed. In short, the assumption of a complete set of correct image features is not valid in practical image understanding.

So one must incorporate *top-down* analysis to find image features missed by the initial segmentation. Such analysis uses relations to generate hypotheses for missing objects: it predicts locations and (appearance) models of missing objects. As discussed in Section 1.3, IUSs need to construct the description of the scene with insufficient information, so that the active hypothesis generation capability of top-down analysis is crucial.

As noted above, the use of relations is very different in the two analysis processes: consistency examination in bottom-up analysis and hypothesis generation in top-down analysis. Although many IUSs incorporate both these analysis processes, they have no unified reasoning scheme for integrating the bottom-up and top-down analyses: they use ad hoc rules to control the analysis. An important characteristic of our evidence accumulation method is that it enables the system to integrate both these analysis processes into a unified reasoning process. The following sections describe the concept, characteristics, and logical foundations of our reasoning process.

2.2.2. Principle of Evidence Accumulation

The world model involves various spatial relations among objects. Suppose we have the following house–road relation:

A road $road_0$ is *along* a house $house_0$ if the predicate $ALONG(road_0, house_0)$ is true.

As discussed in the previous section, there are at least two potential uses of this relation:

1. In the bottom-up analysis, use the relation to check whether road $road_0$ is along house $house_0$.
2. In the top-down analysis, use the relation to direct a search for a road along house $house_0$ and vice versa.

In order to support multiple uses of a relation, we represent a spatial relation by a pair of functional descriptions and use a *hypothesize-and-test paradigm* for spatial reasoning. A binary relation $REL(O_1, O_2)$ between object classes O_1 and O_2 is represented using the following two functional descriptions:

$$O_1 = f(O_2) \quad \text{and} \quad O_2 = g(O_1) \tag{2.1}$$

Given an instance of O_2, say $\mathbf{s}$, function f maps it into a description of an instance of O_1, $f(\mathbf{s})$, which satisfies the relation REL with $\mathbf{s}$: $REL(f(\mathbf{s}), \mathbf{s})$ is true. The analogous interpretation holds for the other function g. In the rest of the book, we call functional descriptions generated by object instances like $f(\mathbf{s})$ *hypotheses,* and *pieces of evidence* refer to object instances or hypotheses.

In a sense, we *compile* relation $REL(O_1, O_2)$ into the functional descriptions of (2.1), and the compiled information is recorded in the world model. That is, we store a pair of procedures representing functions f and g separately in the world model. Note that it is the user's responsibility to preserve the consistency between these functions; function g is the reverse function of f and vice versa [e.g., $\mathbf{s} = g(f(\mathbf{s}))$]. The stored procedures are used to generate hypotheses assuming they are consistent.

As will be described in detail in Section 2.4, the knowledge about a class of objects is represented by an *object class.* We store a group of production rules in each object class, based on which its instance conducts reasoning. Each rule consists of a precondition and an action, and functions like f and g are stored in the action parts of the rules. That is, a precondition represents a set of conditions specifying when the function can be applied, and an action represents the computational procedure corresponding to the function to generate a hypothesis.

Whenever an object instance is created and the conditions are satisfied, the function is applied to generate a *hypothesis* [i.e., $f(\mathbf{s})$] for its related object which would, if found, satisfy the relation with the original

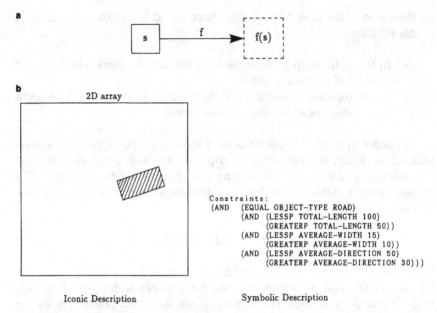

Figure 2.4. Hypothesis generation by an object instance. (a) Hypothesis generation. (b) Description of a hypothesis.

instance (Fig. 2.4a). A hypothesis is described in the following two forms:

1. *Iconic description*: a local area (locational constraint) where the target object instance may be located.
2. *Symbolic description*: a set of constraints on the properties of the target object instance.

Figure 2.4b illustrates an example of a hypothesis. Note that all constraints associated with a hypothesis are on intrinsic properties (i.e., attributes) of the target object. In other words, no geometric constraints are associated with hypotheses. For example, suppose a house instance **a** wants to find a driveway connected to it. **a** generates a hypothesis with which a set of constraints on attributes of the target driveway are associated: location, width, length, and so on. But the geometric constraint that the target driveway must be connected to **a** is not associated with the hypothesis. The reason for this and how much geometric constraints are managed will be discussed in Sections 2.3.5 and 3.9.

All pieces of evidence (i.e., object instances and hypotheses) are stored in the Iconic/Symbolic Database (Fig. 2.1), where accumulation of evidence (i.e., recognition of consistent instances and hypotheses) is performed. This database contains an iconic data structure (i.e., a 2D array for 2D scene analysis) to represent locational constraints associated with the stored pieces of evidence. Each piece of evidence is represented by a 2D region in the array, and the mutual relations of the pieces (i.e., inclusions and overlaps) are structurally represented by a lattice (a tree structure sharing lower and leaf nodes). GRE uses this lattice to index mutually consistent pieces of evidence. Note that this array represents the scene under analysis, so that its coordinate system is defined independently of that of the image. Moreover, we should use a three-dimensional (3D) array for a 3D scene. Besides this locational information, the database records symbolic information, such as established relations among and various attributes of object instances.

Suppose instance **s** of object O_2 generates a hypothesis $f(\mathbf{s})$ for object O_1 based on relation REL and the region associated with $f(\mathbf{s})$ overlaps with an instance of O_1, **t** (Fig. 2.5a). If the set of constraints associated with $f(\mathbf{s})$ is satisfied by **t**, these two pieces of evidence are combined to form what we call a *situation*. That is, a situation is defined by a group of mutually consistent pieces of evidence. GRE *unifies* $f(\mathbf{s})$ and **t**, and establishes the (symbolic) relation REL between **s** and **t** as the result of resolving the situation. More specifically, when GRE succeeds in unifying $f(\mathbf{s})$ and **t**, it reports that result to both **s** and **t**. Then each object

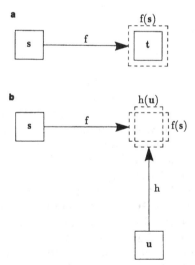

Figure 2.5. Evidence accumulation. (a) Situation which activates the bottom-up analysis. (b) Situation which activates the top-down analysis.

instance records the symbolic relation REL, retracts the related hypothesis [i.e., $f(s)$ for s and $g(t)$ for t if it was generated], and, if so specified, generates other hypotheses based on the stored production rules. (Recall that in SIGMA each object instance itself is an active reasoning agent.) Note that while instance s in the above example has verified the relation to t through hypothesis $f(s)$, t does not verify its reverse relation, believing that it can also be verified; the consistency between pairs of functions like f and g is implicitly assumed. This is the form of a bottom-up analysis to establish a spatial relation between a pair of object instances.

On the other hand, a situation may consist of consistent hypotheses alone (Fig. 2.5b). In this case, their unification leads GRE to search for an instance of the hypothesized object in the image. First GRE unifies all hypotheses included in the situation into a *composite hypothesis*. In this unification, the sets of constraints associated with the hypotheses are integrated by computing their intersection: all constraints are ANDed. Then GRE asks MSE to detect the instance denoted by the composite hypothesis, and MSE in turn activates LLVE. If the instance is successfully detected, it is inserted into the Iconic/Symbolic Database, and the relations between the new instance and the *source* instances, which generated the hypotheses, are established. Here again, GRE reports the new instance to each source instance (i.e., s and u in Fig. 2.5a), which establishes the symbolic relation and retracts the related hypothesis. This is the form of a top-down analysis to find a missing object instance.

The major roles of GRE can be summarized as follows:

1. *Consistency examination.* Examine the consistency among pieces of evidence in the database. Consistency here means that consistent pieces of evidence, such as $f(s)$ and t, denote the same entity.

2. *Unification of consistent evidence.* Integrate multiple different descriptions of the same entity into a unified one. As will be described in the next section, sophisticated unification capability is required in performing spatial reasoning based on PART-OF relations.

3. *Focus of attention.* Many hypotheses are generated and consequently many consistent situations that must be resolved are formed by the consistency examination. GRE has to select the best one to pursue. We call this function of GRE *focus of attention*.

4. *Resolving the selected situation.* Activate either bottom-up or top-down analysis depending on the nature of the selected situation. That is, if the situation includes an object instance, then activate the bottom-up analysis. Otherwise, the top-down analysis is activated.

Recall that hypothesis generation and the process of establishing symbolic relations are conducted by object instances. Thus, GRE is the global coordinator to realize cooperative reasoning by independent reasoning agents (i.e., object instances): GRE determines which agent should communicate (be consistent) with which agents. It should be noted that since hypotheses are just (passive) data structures representing constraints on expected objects (Fig. 2.4), no new hypotheses are generated by hypotheses.

A major characteristic of our process of evidence accumulation is the active generation of explicit descriptions of partial evidence based on knowledge about spatial relations among objects. Consequently, multiple descriptions denoting the same object are generated. Although this may seem to be redundant and large amounts of computation time and storage space are required to manage such descriptions, a robust and flexible spatial reasoning can be realized. That is, our spatial reasoning can be initiated at any time based on the incomplete information thus far obtained, and the accumulation of multiple pieces of evidence from different sources increases the reliability of the analysis; we can construct the description of the scene even if some pieces of evidence are missing. Thus, the redundancy of the system complements the insufficiency of the input information. These merits and demerits of our reasoning method are common to other evidential reasoning schemes like the Hough transform (see Section 1.3). Although we have not studied the computational aspects of our reasoning method (Tsot1988), some discussions will be devoted in Chapter 6 to the feasibility of using parallel reasoning to reduce the computation time.

2.2.3. Reasoning Based on PART-OF Relations

Two types of geometric relations are used in SIGMA: *spatial relations* (SP) and PART-OF *relations* (PO). PO relations are used to construct hierarchical tree structures representing objects with complex internal structures, while SP relations represent geometric relations between different classes of objects. These two types of relations are used differently in spatial reasoning: while the reasoning based on SP relations is performed as described in the previous section, that based on PO relations is slightly different and requires more complicated processing.

This difference comes from the *directionality* involved in PO relations. In the case of an SP relation, a pair of objects linked by such a relation be considered as being at the same level, and no specific directional information is involved in the relation. For instance, relation ALONG between HOUSE and ROAD in the previous example is an SP relation and both objects can use it in the same way. On the other hand, a PO relation links a whole object with its component objects. The levels of these two objects are different. Thus, the PO relation involves directionality: the bottom-up use from the lower level (i.e., the component object) to the higher level (i.e., the whole object) and the top-down use from the higher level to the lower level. We give different semantics to the bottom-up and top-down uses of a PO relation, while an SP relation is used in the same way irrespective of the direction.

Suppose the PO hierarchy illustrated in Fig. 2.6a is given in the world model: object classes represented by nodes are linked by PO relations to form a tree structure. We call objects corresponding to leaf nodes in the tree *primitive objects* and the others *composite objects*. In general, primitive objects are recognized first and object instances representing them are generated, because their appearances are simple and correspond directly to image features.

While instances of primitive objects are represented by both iconic and symbolic descriptions in the database, those of composite objects are

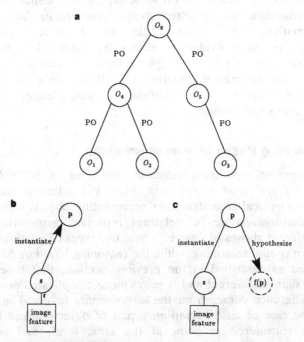

Figure 2.6. Reasoning based on *PART-OF* relations. (a) A *PART-OF* hierarchy. (b) Bottom-up instantiation. (c) Top-down hypothesis generation.

represented only symbolically by instantiated PO hierarchies (i.e., tree structures consisting of object instances). This implies that instances of composite objects refer to abstract groups of primitive object instances and consequently cannot be used directly for evidence accumulation; only pieces of evidence whose iconic descriptions are consistent (i.e., intersecting) are accumulated. On the other hand, hypotheses about composite objects have iconic descriptions, so that they can interact with other pieces of evidence.

The reason why we prohibit instances of composite objects from having iconic descriptions is that their iconic descriptions always overlap with those of their component objects and consequently many redundant situations are generated. In particular, the further the interpretation proceeds, the more redundant situations are generated. Such redundant situations increase the storage space required for the database and the computation time needed for GRE to compute useful situations.

2.2.3.1. Bottom-Up Instantiation of Parent Objects

Let s denote an instance of object O_1 in Fig. 2.6a. It can *directly instantiate* its parent object through the PO relation. That is, in the bottom-up use of a PO relation, an object instance rather than a hypothesis is generated (Fig. 2.6b). The rationale behind this mechanism is as follows. Since a part of an object (i.e., an instance of a primitive object) has been detected, we can consider that the whole object is also detected. For example, if we see a tire, we believe there exists a vehicle. Since the other parts have not been verified yet, the whole object is represented by a partially instantiated PO hierarchy in which many part objects are missing.

Once an instance of object O_4, p, is generated by s, then p in turn may generate an instance of its parent object, O_6 (Fig. 2.6a). This bottom-up instantiation through the PO hierarchy is controlled by a *kernel list* associated with each object class.

An object instance in SIGMA is in one of the following two states: *fully instantiated* or *partially instantiated*. The kernel list is used to discriminate between these two states.

Suppose composite object O is composed of part objects $P_1, P_2, \ldots, P_n$. A kernel list consists of a set of sublists. Each sublist in the kernel list of O specifies a subset of $\{P_1, P_2, \ldots, P_n\}$, where $\{\ldots\}$ denotes a set. Object O is fully instantiated if all part objects in at least one sublist are fully instantiated. Otherwise, it is partially instantiated. Only a fully instantiated object instance can instantiate its parent object via a PO relation. Thus, in the previous example, even if s instantiates its parent object O_4 to generate **p** (Fig. 2.6b), **p** does not instantiate its parent object O_6 when **p** is not fully instantiated. In general, the kernel list of each leaf node object in PO hierarchies is (( )) (i.e., a list including a null sublist). This means that instances of primitive objects are always fully instantiated. In other words, once a primitive object is instantiated, it always instantiates its parent object.

The kernel list can be considered as the decision criterion to decide whether or not enough evidence to support the existence of a composite object instance has been obtained. That is, each sublist specifies a set of part objects which must be recognized to believe the existence of the whole object. In this sense, only fully instantiated object instances are considered as real instances, while partially instantiated instances are hypothetical ones because we still need more evidence (i.e., recognition of other part objects) to verify the existence of the whole object. This is the reason why only fully instantiated instances can generate instances of their parent objects.

2.2.3.2. Top-Down Hypothesis Generation

When a composite object is initiated, it may then generate hypotheses for its missing part objects (Fig. 2.6c). This top-down hypothesis generation is the same as that based on an SP relation and can be performed irrespective of the state of an instance: partially instantiated or fully instantiated.

Note that no instantiated part object generates hypotheses directly for the other part objects at the same level in the PO hierarchy. The reason for this is that since there are many possible geometric relations among part objects, it is more convenient to integrate such relations in their parent object, where n-ary relations among the part objects can be easily specified. In this sense, a composite object in SIGMA is a conceptual one storing various geometric relations among a group of its part objects.

2.2.3.3. Unification of Partially Instantiated PO Hierarchies

The interpretation process of constructing instantiated PO hierarchies (i.e., recognition of composite objects) involves some complications. Consider the following situation (Fig. 2.7a). **s** is an instance of primitive object O_1 in Fig. 2.6a and instantiates its parent object O_4 to generate **p**, which then generates a hypothesis for its missing part object O_2, $f(\mathbf{p})$. Suppose an instance of O_2 corresponding to the missing part, say **t**, has already been detected. Then, since the two pieces of evidence

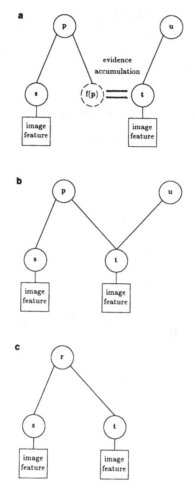

Figure 2.7. Unification of partially instantiated *PART-OF* hierarchies (I). (a) Partially instantiated *PART-OF* hierarchies. (b) Object instance with multiple parents. (c) Unified *PART-OF* hierarchy.

$f(\mathbf{p})$ and **t** are consistent, GRE unifies them to establish a PO relation between **p** and **t**.

However, since **t** is an instance of O_2, it also has instantiated its own parent object O_4 before the unification. Let **u** denote such a parent instance: a pair of instances **t** and **u** have therefore been connected by a PO relation. Thus, as a result of unifying $f(\mathbf{p})$ and **t**, instance **t** comes to have two parent instances, **p** and **u**, at the same time (Fig. 2.7b).

This leads GRE to another unification. When instance **t** finds that it has two parent instances, it asks GRE to unify them. GRE examines the pair of parent instances **p** and **u** and, if they are consistent, it unifies them to generate a new instance of O_4, say **r**, by which a merged instantiated PO hierarchy is constructed (Fig. 2.7c). Note that the unification of multiple parent instances may trigger still another unification of grandparent instances, if such instances have been generated. After the unification of **p** and **u**, the merged parent instance **r** contains both **s** and **t** as its parts. Therefore it can use the properties of these instances to generate a new hypothesis for a still missing part object whose properties could not be specified previously due to a lack of sufficient information.

In short, in our evidence accumulation method, each instance of primitive objects instantiates its parent object and constructs a partially instantiated PO hierarchy independently of the others. This causes multiple descriptions of the same composite (parent) object. A major role of GRE is to unify such duplicated descriptions. The unification of partially instantiated PO hierarchies is triggered by an instance shared by multiple parent instances. By iterating this unification, a completely instantiated PO hierarchy representing a complex composite object is gradually constructed.

If a pair of parent instances sharing the same part instance are found not to be consistent, GRE regards them as conflicting interpretations: a copy of the shared part instance is generated to separate a pair of PO hierarchies, which are marked as mutually conflicting. GRE performs reasoning separately based on each interpretation: no pieces of evidence

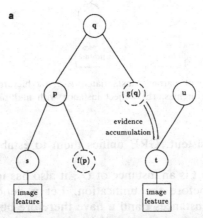

Figure 2.8. Unification of partially instantiated *PART-OF* hierarchies (II). (a) Accumulation of evidence of different object classes. (b) Object instance with multiple parents. (c) Unified *PART-OF* hierarchy. (d) Instantiation of a chain of object classes.

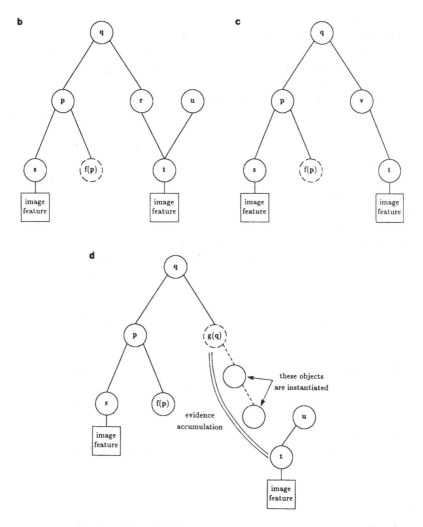

Figure 2.8. (*continued*)

generated from conflicting interpretations are not accumulated. That is, GRE constructs multiple mutually conflicting interpretations in the database. Some discussion on the reasoning behind the generation of these multiple alternative interpretations will be given in Section 2.3.6.

Still more complicated processing is required to resolve situations formed by reasoning based on PO relations. As illustrated in Fig. 2.8a, suppose the grandparent object O_6 in Fig. 2.6a has also been instantiated by the parent object instance **p**. Let **q** denote this instance. In this case, **q** as well as **p** generates a hypothesis for its missing part. Let $g(\mathbf{q})$ denote the hypothesis for composite object O_5 in Fig. 2.6a. Suppose as before, an instance of O_3, **t**, has been detected and has generated an instance of O_5, **u**. If a pair of instances **s** and **t** correspond to parts of the same composite object, then hypothesis $g(\mathbf{q})$ and instance **t** are consistent and hence accumulated to form a situation. It should be noted that although $g(\mathbf{q})$ and **u** are denoting the same object, they are not directly accumulated, because **u** has no iconic description.

In the case shown in Fig. 2.8a, the object classes to which the accumulated pieces of evidence belong are at different levels of the PO hierarchy: $g(\mathbf{q})$ belongs to O_5 and **t** to O_3. Thus, in forming the situation consisting of $g(\mathbf{q})$ and **t**, GRE examines their consistency by checking if **t** can be a part of $g(\mathbf{q})$. The consistency examination by GRE involves this type of processing as well as the process of identifying the equality between pieces of evidence generated from different sources.

In resolving the situation consisting of $g(\mathbf{q})$ and **t**, GRE first generates a new instance of composite object O_5, say **r**, and establishes a PO relation between **r** and **t** (Fig. 2.8b). Here again instance **t** is shared by two different parents, **r** and **u**. Then **t** asks GRE to unify these two parents. By this unification, a new instance of composite object O_5, **v**, is generated, and all instances **s**, **p**, **q**, **v**, and **t** are organized into one PO hierarchy (Fig. 2.8c).

In general, the process of resolving a situation which consists of pieces of evidence at different levels of a PO hierarchy requires the instantiation of new composite objects. As illustrated in Fig. 2.8d, for example, if two pieces of consistent evidence $g(\mathbf{q})$ and **t** belong to object classes at very distant levels in a PO hierarchy, GRE instantiates a chain of ancestor objects from **t** to generate an instance at the same level as $g(\mathbf{q})$.

2.2.3.4. Summary

In summary, GRE has to perform the following functions to realize spatial reasoning based on PO relations:

1. Unification of partially instantiated PO hierarchies
2. Consistency examination between pieces of evidence at different levels of a PO hierarchy
3. Instantiation of intermediate composite objects to unify a pair of partially instantiated PO hierarchies

Functions (2) and (3) would not be required if we prohibited the

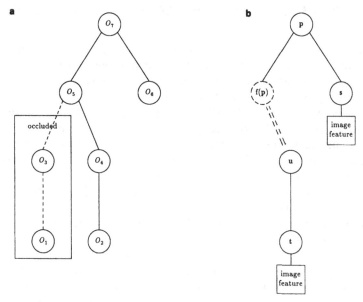

Figure 2.9. Coping with occlusion. (a) *PART-OF* hierarchy with occlusion. (b) Partially instantiated *PART-OF* hierarchies.

accumulation of evidence belonging to object classes at different levels of a PO hierarchy. With such a reasoning mechanism, however, we would encounter the following problem. Suppose a part of a composite object is occluded, as shown in Fig. 2.9a, and a pair of partially instantiated PO hierarchies such as those in Fig. 2.9b are constructed. That is, s and t are instances of primitive objects O_6 and O_2, respectively. s instantiated its parent object O_7 to generate **p**, which then generated a hypothesis for O_5, $f(\mathbf{p})$. Similarly, t generated its parent object O_4 to generate **u**. In this case, if we did not allow the accumulation of $f(\mathbf{p})$ and t, no further reasoning could be performed. Note that since **u** has no iconic description, it cannot be accumulated with other pieces of evidence. Note also that hypothesis $f(\mathbf{p})$ cannot generate a new hypothesis for object O_4 in Fig. 2.9a to which instance **u** belongs; if hypotheses were allowed to generate hypotheses, infinitely many hypotheses would be generated based on uncertain information. Thus to cope with occlusion, we need to accumulate pieces of evidence even if they belong to object classes at different levels of a PO hierarchy. In our method, hypothesis $f(\mathbf{p})$ and instance t in Fig. 2.9b can be accumulated, and the pair of partially instantiated PO hierarchies are unified to form an integrated PO hierarchy.

2.3. LOGICAL FOUNDATIONS OF EVIDENCE ACCUMULATION

Although many reasoning methods for image understanding have been proposed, few of them have precise definitions; verbal descriptions and illustrative examples have been used to explain how the reasoning is performed, just as we did in the previous section. As Reiter and Mackworth pointed out (Reit1987), formal logic allows us to precisely describe and discuss the knowledge and reasoning mechanisms required in image understanding: discrimination between scene domain knowledge and image domain knowledge, the precise specification of the task of image understanding, and what interpretation means in image understanding. In this section, we try to formulate the foundations of our reasoning method using the first-order predicate calculus.

In Chapter 1, we emphasized that the interpretation process of image understanding systems should be constructive: it should actively generate new information based on the world model to cope with the insufficiency of the input information. We proposed the evidence accumulation method to realize such constructive reasoning. The major purposes of this section are to discuss its logical foundation in terms of the first-order predicate calculus, to clarify the difference between our reasoning method and ordinary deductive reasoning (i.e., theorem proving), and accordingly to disclose the reason why we call our reasoning constructive.

In what follows, we use the following logical symbols:

$\forall$	Universal quantifier
$\exists$	Existential quantifier
$\land$	Logical *AND*
$\lor$	Logical *OR*
$\supset$	Logical *IMPLICATION*
$\neg$	Logical *NOT*

We describe predicate symbols in capital letters, variables and function symbols in lowercase letters, and constants in bold letters. Since the logic we will use is the ordinary first-order predicate calculus, we do not explain how logical formulas are described in terms of these symbols or inference rules in the first-order predicate calculus (e.g., the *resolution principle*); see Mend1964 for a formal definition of the first-order predicate calculus, Chan1973 for various reasoning methods based on the resolution principle, and Nils1980 and Gene1987 for the use of the logic in artificial intelligence (AI).

In general, we have to use a pair of disjoint sets of constants to denote a set of image features and a set of scene objects in image understanding; these two types of entities must be clearly separated and what we call interpretation is the establishment of correspondence between these two sets of entities (see Fig. 1.1). However, since major discussions in this section deal with spatial reasoning based on scene domain knowledge, all constants and variables denote objects in the scene. In Section 2.3.5, where we introduce constants denoting image features, we will discuss the logical framework of the entire image understanding process in SIGMA.

2.3.1. Knowledge Representation and Hypothesis Generation in Logic

Suppose we want to describe the following knowledge in terms of logical formulas:

A spatial relation *REL* (e.g., *ALONG*) holds

between *HOUSE* and *ROAD*. (2.2)

In terms of the first-order predicate calculus, this knowledge can be described as

$$\forall x[HOUSE(x) \supset \exists y\{ROAD(y) \wedge REL(x, y)\}] \qquad (2.3)$$

where $HOUSE(x)$ and $ROAD(y)$ are predicates which specify object classes: for example, if constant **a** denotes a house instance, $HOUSE(\mathbf{a})$ becomes *TRUE*. This axiom can read as "for every house, there exists a road which satisfies the relation *REL* to it." In general, the knowledge can be described by using (1) unary predicates specifying object classes and (2) *n*-ary predicates specifying spatial relations. As will be discussed in Section 2.3.5, these two types of information are processed differently in SIGMA. Note that most SP and PO relations used in SIGMA are binary relations and that *n*-ary relations are implicitly used in the reasoning by object instances (see the discussion at the end of this section).

Although axioms like (2.3) seem to be natural to describe knowledge about spatial relations among objects, we should note the following points:

1. *Uniqueness constraint.* Since *REL* represents a spatial relation, a unique road instance satisfies the relation to each house instance; since *REL* specifies the relative location from a house instance to its related road

instance, multiple different road instances cannot be located at the same location. However, axiom (2.3) involves no information about this uniqueness constraint. To introduce the uniqueness constraint, (2.3) is to be extended to

$$\forall x[HOUSE(x) \supset \exists y[ROAD(y) \land REL(x, y) \land \forall z\{ROAD(z)$$
$$\land REL(x, y) \supset z = y\}]] \qquad (2.4)$$

That is, we need to introduce the equality relation to represent the uniqueness constraint. As will be discussed later, the equality relation plays a crucial role in our reasoning.

2. *Directionality of a relation.* Although the knowledge in (2.2) (i.e., the relation) has no directionality, axiom (2.3) [and (2.4)] involves the implicit direction from *HOUSE* to *ROAD*. In other words, the reasoner can perform reasoning based on this knowledge only if some information about *HOUSE* is given. For example, suppose the reasoner uses *modus ponens* as the inference rule. Then, given the fact *HOUSE*(**a**) is *TRUE*, one can deduce

$$\exists y[ROAD(y) \land REL(\mathbf{a}, y) \land \forall z\{ROAD(z) \land REL(\mathbf{a}, z) \supset z = y\}] \qquad (2.5)$$

based on axiom (2.4). On the other hand, the reasoner cannot perform any reasoning even if *ROAD*(**b**) is given as a fact. Hence the following *dual* axiom should also be used to describe the original knowledge:

$$\forall x[ROAD(x) \supset \exists y[HOUSE(y) \land REL(y, x) \land \forall z\{HOUSE(z)$$
$$\land REL(z, x) \supset z = y\}]] \qquad (2.6)$$

Thus, the knowledge in (2.2) is described by axioms (2.4) and (2.6).

Since we will use the resolution principle as the inference rule, we transform these axioms into the following sets of *clauses* (note that we do not use proof by refutation, which is a popular deductive reasoning scheme):

Clauses from (2.4):

$$\neg HOUSE(x) \lor ROAD(f(x)) \qquad (2.7a)$$

$$\neg HOUSE(x) \lor REL(x, f(x)) \qquad (2.7b)$$

$$\neg HOUSE(x) \lor \neg ROAD(z) \lor \neg REL(x, z) \lor z = f(x) \qquad (2.7c)$$

Clauses from (2.6):

$$\neg ROAD(x) \lor HOUSE(g(x)) \tag{2.8a}$$

$$\neg ROAD(x) \lor REL(g(x), x) \tag{2.8b}$$

$$\neg ROAD(x) \lor \neg HOUSE(z) \lor \neg REL(z, x) \lor z = g(x) \tag{2.8c}$$

It should be noted that (1) variables in different clauses are independent (different) even if the same variable symbol is used, (2) all variables are universally quantified, and (3) all clauses are logically connected by $\land$.

Functions $f(x)$ and $g(x)$ are called *Skolem functions*, and have been introduced to remove the existential quantifiers from the axioms. $f(x)$ denotes a certain object which satisfies the relation REL to object x (i.e., $REL(x, f(x))$ is $TRUE$ if $HOUSE(x)$ holds), and the meaning of the function is to map object x to such an object.

Suppose that $HOUSE(\mathbf{a})$ is given as a fact: a house instance $\mathbf{a}$ is recognized. Then, the following three new clauses are deduced from (2.7):

$$ROAD(f(\mathbf{a})) \tag{2.9a}$$

$$REL(\mathbf{a}, f(\mathbf{a})) \tag{2.9b}$$

$$\neg ROAD(z) \lor \neg REL(\mathbf{a}, z) \lor z = f(\mathbf{a}) \tag{2.9c}$$

Here $f(\mathbf{a})$ denotes a certain unknown constant (object instance), and its properties are specified by the derived clauses (2.9a) and (2.9b): $f(\mathbf{a})$ must be an instance of $ROAD$ and must satisfy relation REL to $\mathbf{a}$. In other words, we can consider these derived clauses as constraints on the properties of $f(\mathbf{a})$.

Such constraints can be specified more precisely if we have detailed knowledge about $ROAD$ and REL:

$$\forall x[ROAD(x) \supset GREATER(\textit{width-of}(x), 5)$$

$$\land LESS(\textit{width-of}(x), 100) \land RIBBON(\textit{shape-of}(x))] \tag{2.10}$$

$$\forall x, y[REL(x, y) \supset PARALLEL(\textit{axis-of}(x), \textit{axis-of}(y))$$

$$\land DISTANCE(\textit{centroid-of}(x), \textit{centroid-of}(y), 50)], \tag{2.11}$$

where *width-of, shape-of, axis-of,* and *centroid-of* denote functions to compute attributes of objects and their semantics are defined by *attached procedures* (Nils1980). That is, these functions are not *functors* to construct data structures (e.g., *cons* for list structures) but can be considered as real functions with computational capabilities. Based on these axioms, more detailed constraints on the properties of $f(\mathbf{a})$ can be derived. For example, using (2.9a) and (2.10), we can derive the following constraint:

$$GREATER(width\text{-}of(f(\mathbf{a})), 5) \wedge LESS(width\text{-}of(f(\mathbf{a})), 100)$$

$$\wedge \ RIBBON(shape\text{-}of(f(\mathbf{a}))) \qquad (2.12)$$

As is obvious from the above discussion, this reasoning process is exactly the same as our hypothesis generation process. Our object instance corresponds to a constant in the logic, which generates a hypothesis for its related object by applying a function to itself. The generated hypothesis corresponds to an instantiated Skolem function (i.e., $f(\mathbf{a})$) in the logic, with which a set of constraints are associated. That is, axioms (2.4), (2.10), and (2.11) are stored in the object class representing *HOUSE*, based on which the hypothesis and constraints are derived.

From a theoretical point of view, we can derive

$$HOUSE(g(f(\mathbf{a}))) \qquad (2.13)$$

from (2.8a) and (2.9a). In this derivation, a new instantiated Skolem function $g(f(\mathbf{a}))$ is generated. And then, from (2.13) and (2.7a), we have

$$ROAD(f(g(f(\mathbf{a})))) \qquad (2.14)$$

This derivation continues infinitely, and infinitely many instantiated Skolem functions are generated. In our interpretation, $g(f(\mathbf{a}))$ means the hypothesis generated by hypothesis $f(\mathbf{a})$. In SIGMA, in order to avoid such infinite generation of instantiated Skolem functions, we prohibit a hypothesis from generating a new hypothesis. In other words, the terms manipulated by GRE are (in principle) confined to a finite set of ground terms such as constants and nonnested ground (instantiated) functions. As will be described in Section 2.3.4, we allow the generation of nested instantiated Skolem functions in the reasoning based on PO relations.

Note that the reasoning for the hypothesis generation is local; given a constant, the reasoning can be performed independently without using the other constants. This corresponds to our idea of considering object instances as active reasoning agents; they can generate hypotheses based on their own knowledge.

Suppose we have the following knowledge about ternary relation *BETWEEN* among *HOUSE*, *ROAD*, and *D-WAY* (driveway):

$$\forall x[HOUSE(x) \supset \exists y, z[ROAD(y) \wedge D\text{-}WAY(z) \wedge BETWEEN(z, x, y)$$
$$\wedge \; \forall u\{D\text{-}WAY(u) \wedge BETWEEN(u, x, y) \supset u = z\}]] \qquad (2.15)$$

Here we consider that a *HOUSE* may have access to multiple *ROADS*, and be connected to each via a unique *D-WAY*. This axiom is stored in object class *HOUSE* and when its instance **a** is recognized, we can derive the following clauses:

$$ROAD(f(\mathbf{a})) \qquad\qquad\qquad (2.16)$$

$$D\text{-}WAY(g(\mathbf{a})) \qquad\qquad\qquad (2.16b)$$

$$BETWEEN(g(\mathbf{a}), \mathbf{a}, f(\mathbf{a})) \qquad\qquad (2.16c)$$

$$\neg D\text{-}WAY(u) \vee \neg BETWEEN(u, \mathbf{a}, f(\mathbf{a})) \vee u = g(\mathbf{a}) \qquad (2.16d)$$

In this derivation, two instantiated Skolem functions are obtained. This means that house instance **a** generates two hypotheses for *ROAD* and *D-WAY* based on ternary relation *BETWEEN*. In practice, however, hypothesis generation is controlled by the control knowledge stored in the object class. That is, production rules stored in *HOUSE* may generate the hypothesis for *ROAD* alone because the precise location of *D-WAY* can be estimated only if a house instance and its facing road instance are detected.

2.3.2. Reasoning in the Bottom-Up Analysis

When *ROAD*(**b**) as well as *HOUSE*(**a**) is given as another fact, a hypothesis for a house instance denoted by $g(\mathbf{b})$ is generated based on clauses (2.8a) and (2.8b). Then, using this fact and (2.9c), the following new clause is derived:

$$\neg REL(\mathbf{a}, \mathbf{b}) \vee \mathbf{b} = f(\mathbf{a}) \qquad\qquad (2.17)$$

Note that the following dual clause is also derived from (2.8c):

$$\neg REL(\mathbf{a}, \mathbf{b}) \vee \mathbf{a} = g(\mathbf{b}) \qquad\qquad (2.18)$$

Suppose we examine the relation between **a** and **b** and verify that *REL*(**a**, **b**) holds. Then, using this new fact, we can derive $\mathbf{b} = f(\mathbf{a})$ and $\mathbf{a} = g(\mathbf{b})$ from (2.17) and (2.18), respectively. This can be considered as the reasoning in the ordinary bottom-up analysis (see Fig. 2.3).

The direction of the reasoning in our evidence accumulation is the opposite. That is, instead of verifying $REL(\mathbf{a}, \mathbf{b})$, we examine $\mathbf{b} = f(\mathbf{a})$ and/or $\mathbf{a} = g(\mathbf{b})$ to derive $REL(\mathbf{a}, \mathbf{b})$ as the conclusion; GRE in SIGMA examines the consistency between $\mathbf{b}$ and $f(\mathbf{a})$ (or $\mathbf{a}$ and $g(\mathbf{b})$) to establish the relation REL between $\mathbf{a}$ and $\mathbf{b}$. Suppose we verify that $\mathbf{b} = f(\mathbf{a})$ holds: that is GRE regards $\mathbf{b}$ and $f(\mathbf{a})$ as denoting the same object. Then, we can derive $REL(\mathbf{a}, \mathbf{b})$ from (2.9b) by substituting $f(\mathbf{a})$ with $\mathbf{b}$. (The equality axioms, which describe the properties of the special predicate $=$, justify this substitution.) Since now we have $REL(\mathbf{a}, \mathbf{b})$, we can derive $\mathbf{a} = g(\mathbf{b})$ from (2.18).

More precisely speaking, when GRE verifies $\mathbf{b} = f(\mathbf{a})$, it reports this fact to object instance $\mathbf{a}$, which derives $REL(\mathbf{a}, \mathbf{b})$ based on its local knowledge. Then, $\mathbf{a}$ reports $REL(\mathbf{a}, \mathbf{b})$ to $\mathbf{b}$, which in turn derives $\mathbf{a} = g(\mathbf{b})$ based on its own knowledge. Thus, a pair of local reasoning processes conducted by $\mathbf{a}$ and $\mathbf{b}$ are coordinated by GRE. Note that even if we first verify $\mathbf{a} = g(\mathbf{b})$ rather than $\mathbf{b} = f(\mathbf{a})$, we can derive the same conclusions. This demonstrates the flexibility of our reasoning; the reasoning can be initiated from either $\mathbf{a}$ or $\mathbf{b}$. This is the reasoning performed in bottom-up analysis in SIGMA.

In summary,

1. The consistency examination by GRE implies the identification of object instances denoted in different ways (e.g., $\mathbf{a}$ and $g(\mathbf{b})$). In other words, since we admit that the same object can be denoted in many different ways, the system needs to have the capability to identify the denoted object. This capability is called *semantic matching* in AI (Nils1980), and the performance of the system is heavily dependent on it. Thus, the kernel function of GRE is consistency examination, which introduces equality relations between ground terms (i.e., constants and instantiated functions).

2. As is obvious from the above discussion, our reasoning is not goal-directed as in Prolog (Cloc1981) and ordinary theorem proving. We do not use any goal in the above derivation. Our implicit goal is to establish relations (i.e., derive theorems like $REL(\mathbf{a}, \mathbf{b})$), a process which corresponds to constructing the structural description of the scene. From a computational point of view, however, data-directed (forward) reasoning like ours requires a reasoning strategy to reduce the amount of computation; since a huge number of theorems can be derived in straightforward data-driven reasoning, a criterion to determine which theorems are to be derived is required to reduce the number of derived theorems. It is for this purpose that we prohibit the

generation of nested instantiated Skolem functions (i.e., prohibit a hypothesis from generating another new hypothesis). The *focus of attention mechanism* is introduced in GRE to guide the reasoning process and hence reduce the number of derived theorems. We will discuss this problem in Chapter 3.

2.3.3. Reasoning in the Top-Down Analysis

Suppose we are given the following axioms as the knowledge about *HOUSE*, *ROAD*, and *D-WAY*:

$$\forall x[HOUSE(x) \supset \exists y[D\text{-}WAY(y) \wedge REL1(x, y) \wedge \forall z\{D\text{-}WAY(z)$$
$$\wedge REL1(x, z) \supset z = y\}]] \quad (2.19)$$
$$\forall x[ROAD(x) \supset \exists y[D\text{-}WAY(y) \wedge REL2(x, y) \wedge \forall z\{D\text{-}WAY(z)$$
$$\wedge REL2(x, z) \supset z = y\}]] \quad (2.20)$$

The following clauses are obtained from these axioms:

Clauses from (2.19):

$$\neg HOUSE(x) \vee D\text{-}WAY(f(x)) \quad (2.21a)$$
$$\neg HOUSE(x) \vee REL1(x, f(x)) \quad (2.21b)$$
$$\neg HOUSE(x) \vee \neg D\text{-}WAY(z) \vee \neg REL1(x, z) \vee z = f(x) \quad (2.21c)$$

Clauses from (2.20):

$$\neg ROAD(x) \vee D\text{-}WAY(g(x)) \quad (2.22a)$$
$$\neg ROAD(x) \vee REL2(x, g(x)) \quad (2.22b)$$
$$\neg ROAD(x) \vee \neg D\text{-}WAY(z) \vee \neg REL2(x, z) \vee z = g(x) \quad (2.22c)$$

When we are given a house instance **a** and a road instance **b**, that is, *HOUSE*(**a**) and *ROAD*(**b**) hold, the following clauses can be derived from (2.21) and (2.22):

$$D\text{-}WAY(f(\mathbf{a})) \quad (2.23a)$$
$$REL1(\mathbf{a}, f(\mathbf{a})) \quad (2.23b)$$
$$\neg D\text{-}WAY(z) \vee \neg REL1(\mathbf{a}, z) \vee z = f(\mathbf{a}) \quad (2.23c)$$
$$D\text{-}WAY(g(\mathbf{b})) \quad (2.23d)$$
$$REL2(\mathbf{b}, g(\mathbf{b})) \quad (2.23e)$$
$$\neg D\text{-}WAY(z) \vee \neg REL2(\mathbf{b}, z) \vee z = g(\mathbf{b}) \quad (2.23f)$$

Since $f(\mathbf{a})$ and $g(\mathbf{b})$ belong to D-WAY [from (2.23a) and (2.23d)], they can be denoting the same object instance. So GRE examines their consistency, and if they are consistent, we can obtain a new fact $f(\mathbf{a}) = g(\mathbf{b})$. Based on this new fact, we can derive new facts representing integrated properties of $f(\mathbf{a})$ and $g(\mathbf{b})$: the constraints on the properties of these instantiated functions are integrated [see (2.10) and (2.11)]. These integrated constraints correspond to our composite hypothesis in SIGMA.

However, the reasoning stops at this stage, because we do not know the constant denoting the same object as $f(\mathbf{a})$ and $g(\mathbf{b})$. Then GRE asks MSE (and LLVE) to find an object instance (i.e., constant) which is denoted by (i.e., satisfies the constraints associated with) both $f(\mathbf{a})$ and $g(\mathbf{b})$. Suppose such instance is detected and is represented by constant $\mathbf{c}$. That is, MSE introduces a new fact D-WAY($\mathbf{c}$) and GRE establishes the equality relation $\mathbf{c} = f(\mathbf{a}) = g(\mathbf{b})$. Then we can derive the following new facts about relations:

$$REL1(\mathbf{a}, \mathbf{c}) \tag{2.24}$$

$$REL2(\mathbf{b}, \mathbf{c}) \tag{2.25}$$

Here again GRE reports $\mathbf{c} = f(\mathbf{a})$ and $\mathbf{c} = g(\mathbf{b})$ to $\mathbf{a}$ and $\mathbf{b}$ respectively, which then derive the new facts (2.24) and (2.25) based on their own local knowledge, respectively. This is the reasoning done in our top-down analysis.

2.3.4. Reasoning in the Construction of PART-OF Hierarchies

As described in Section 2.2.3, spatial reasoning based on PO relations requires more complicated processing. Consider the following knowledge:

A HOUSE-GROUP is an ordered group of HOUSEs and
each pair of successive member HOUSEs satisfies
spatial relation REL. $\tag{2.26}$

In order to represent this knowledge in terms of the logic, we introduce predicate PO(x, y), which means x is a part of y. Then, first we have the following axiom:

$$\forall x[HOUSE(x) \supset \exists y[HOUSE\text{-}GROUP(y) \land PO(x, y)$$
$$\land \ \forall z\{HOUSE\text{-}GROUP(z) \land PO(x, y) \supset z = y\}]] \tag{2.27}$$

This implies that for every house, there exists a unique house-group to which it belongs. Note that the form of (2.27) is exactly the same as that of (2.4) [and (2.6)]. Thus, when $HOUSE(\mathbf{a})$ is given as a fact, we can perform the same reasoning as in Section 2.2.1 for generating a hypothesis: $f(\mathbf{a})$ is created, which satisfies

$$HOUSE\text{-}GROUP(f(\mathbf{a})) \tag{2.28a}$$

$$PO(\mathbf{a}, f(\mathbf{a})) \tag{2.28b}$$

$$\neg HOUSE\text{-}GROUP(z) \vee \neg PO(\mathbf{a}, z) \vee z = f(\mathbf{a}) \tag{2.28c}$$

Thus, an instance of a composite object generated by the bottom-up use of a PO relation (see Fig. 2.6) corresponds to an instantiated Skolem function, which is the same as a hypothesis generated by an SP relation.

Axiom (2.27) represents only partial information included in the knowledge (2.26). We need to define the properties of $HOUSE\text{-}GROUP$:

$$\forall x[HOUSE\text{-}GROUP(x) \supset \forall x[HOUSE(y) \wedge PO(y, x)$$
$$\supset \exists z\{HOUSE(z) \wedge REL(y, z) \wedge PO(z, x) \wedge \forall w\{HOUSE(w)$$
$$\wedge REL(\mathbf{y}, \mathbf{w}) \supset z = w]\}]] \tag{2.29}$$

Axiom (2.29) describes the structure of the composite object: the relation between member $HOUSES$ in a $HOUSE\text{-}GROUP$.

When we transform (2.29) into clauses, variable z is replaced by a Skolem function $g(x, y)$, because it is included in the scopes of two universally quantified variables x and y. This implies that properties of a member $HOUSE$ in a $HOUSE\text{-}GROUP$ depend on both those of its neighboring $HOUSE$ and the $HOUSE\text{-}GROUP$. Then, as was done before, when $HOUSE(\mathbf{a})$ is given as a fact, we can derive the following facts based on (2.29):

$$HOUSE(g(f(\mathbf{a}), \mathbf{a})) \tag{2.30a}$$

$$REL(\mathbf{a}, g(f(\mathbf{a}), \mathbf{a})) \tag{2.30b}$$

$$PO(g(f(\mathbf{a}), \mathbf{a}), f(\mathbf{a})) \tag{2.30c}$$

$$\neg HOUSE(w) \vee \neg REL(\mathbf{a}, w) \vee g(f(\mathbf{a}), \mathbf{a}) = w \tag{2.30d}$$

Here $g(f(\mathbf{a}), \mathbf{a})$ corresponds to our hypothesis generated by top-down hypothesis generation based on a PO relation (see Fig. 2.6), with which a set of constraints [i.e., (2.30a)–(2.30c)] are associated.

According to our original interpretation of Skolem functions, $g(f(\mathbf{a}), \mathbf{a})$ is generated by hypothesis $f(\mathbf{a})$ and instance $\mathbf{a}$. As discussed before, in principle we prohibit a hypothesis from generating another hypothesis, because too many hypotheses would be generated based on uncertain evidence. However, since we want to perform reasoning based on such axioms as (2.29) representing structures of composite objects, we regard hypotheses generated by the bottom-up use of PO relations [e.g., $f(\mathbf{a})$] as *quasi*-instances, and allow them to generate hypotheses like $g(f(\mathbf{a}), \mathbf{a})$. The same arguments holds in the instantiation of composite objects at higher levels in a PO hierarchy. For example, the grandparent object instance of $\mathbf{a}$ is denoted by $f(f(\mathbf{a}))$, and its generation is controlled by the kernel list associated with the parent object: quasi-instance $f(\mathbf{a})$ decides based on the kernel list if it should generate $f(f(\mathbf{a}))$.

Suppose another *HOUSE* instance $\mathbf{b}$ is given, i.e., *HOUSE*$(\mathbf{b})$ holds. Then, the same reasoning as above is performed to derive clauses similar to (2.28) and (2.30):

$$HOUSE\text{-}GROUP(f(\mathbf{b})) \tag{2.31a}$$

$$PO(\mathbf{b}, f(\mathbf{b})) \tag{2.31b}$$

$$\neg HOUSE\text{-}GROUP(z) \vee \neg PO(\mathbf{b}, z) \vee z = f(\mathbf{b}) \tag{2.31c}$$

$$HOUSE(g(f(\mathbf{b}), \mathbf{b})) \tag{2.31d}$$

$$REL(\mathbf{b}, g(f(\mathbf{b}), \mathbf{b})) \tag{2.31e}$$

$$PO(g(f(\mathbf{b}), \mathbf{b}), f(\mathbf{b})) \tag{2.31f}$$

$$\neg HOUSE(w) \vee \neg REL(\mathbf{b}, w) \vee g(f(\mathbf{b}), \mathbf{b}) = w \tag{2.31g}$$

Note that another quasi-instance of *HOUSE-GROUP* $f(\mathbf{b})$ is generated.

Suppose $\mathbf{a}$ and $\mathbf{b}$ are neighboring *HOUSES* belonging to the same *HOUSE-GROUP*. Then, hypothesis $g(f(\mathbf{a}), \mathbf{a})$ and instance $\mathbf{b}$ are consistent, so that a new fact

$$g(f(\mathbf{a}), \mathbf{a}) = \mathbf{b} \tag{2.32}$$

is introduced by GRE. Based on this fact and (2.30c),

$$PO(\mathbf{b}, f(\mathbf{a})) \tag{2.33}$$

is derived. Then, from (2.28a), (2.31c), and (2.33), we can derive

$$f(\mathbf{a}) = f(\mathbf{b}) \tag{2.34}$$

This corresponds to the unification of a pair of parent instances sharing the same part instance (see Fig. 2.7). Using (2.32) and (2.34), the clauses in (2.28), (2.30), and (2.31) are simplified: duplicated clauses such as (2.28a) and (2.31a) are merged into one clause and $g(f(a), a)$ is replaced by **b**. This simplification process corresponds to the unification of partially instantiated PO hierarchies in SIGMA.

In summary, the most significant distinguishing characteristics in our reasoning are:

1. *Dynamic creation of new facts.* GRE maintains a (finite) set of ground terms [i.e., constants and instantiated functions like **a** and $f(a)$] and examines the consistency among them to find a group of ground terms denoting the same object. Then it generates new facts describing equality relations between such ground terms. MSE, on the other hand, generates new constants and facts about their object classes like $HOUSE(a)$. In SIGMA, these two types of facts are dynamically created during the reasoning.

2. *Dynamic derivation of relational descriptions.* In SIGMA, relational descriptions between objects such as $REL(a, b)$ are not given as facts *a priori* but are dynamically derived from the given axioms as new facts are generated by GRE and MSE. This derivation process corresponds to that of constructing the scene description.

We represented the knowledge about spatial relations in terms of the first-order predicate calculus with equality. Although this formalization is the same as that used in Reit1987 (in fact, most of the axioms used here are similar to theirs), there is a critical difference in the reasoning scheme; ours is constructive while theirs is interpretive. The difference can be clearly understood by comparing the above characteristics of our reasoning with the following principles used in Reit1987:

1. *Unique name and domain closure assumptions* (Reit1980, Gene1987). A set of ground terms is fixed and each term denotes a different entity. In other words, they assume all image features corresponding to objects are given *a priori* and no new image features are introduced during the reasoning.

2. *Axioms as constraints.* Axioms representing the knowledge are regarded as constraints (filters) to obtain possible interpretations (see Fig. 2.3). This paradigm is nothing but the process of relaxation labeling and in fact they mention that the constraint propagation can be used to find interpretations.

2.3.5. Hypothesis-Based Reasoning as a Framework for Image Understanding

Reasoning methods in classic AI systems are based on ordinary deductive reasoning, in which it is assumed that all necessary information is given *a priori* (Nils1980). However, in developing practical AI systems, it is almost impossible to prepare such complete information; the world is too complicated to describe completely in terms of a finite set of axioms. Moreover, even with complete information, there are too many things to be proved to attain a goal in deductive reasoning. For example, suppose a mobile robot is going to cross a street. It has to prove that the traffic signal is green, that all vehicles stop, that the street is not collapsing due to an earthquake, that nothing is falling onto the street, and so on, if it wants to verify the safety of the crossing completely. On the other hand, we human beings seem to perform reasoning moderately well based on *common sense* without caring about very rare exceptional cases such as earthquakes and falling objects.

In AI, several methods have been proposed to formulate reasoning with incomplete knowledge and so-called *common sense reasoning*. In general, they are called *nonmonotonic reasoning* (Spec1980, Gene1987). Recently, Poole et al. (Pool1987) proposed *hypothesis-based reasoning* as a logical framework for nonmonotonic reasoning, expert systems for diagnosis, and learning. Similar reasoning schemes have been proposed by Cox and Pietrzykowski (Cox1986) and Finger and Genesereth (Fing1985). Figure 2.10 illustrates the schematic diagram of their reasoning scheme. Axioms stand for the given knowledge about the world, observations represents facts obtained from the world, and possible logical hypotheses imply various hypothetical axioms which may hold in the world. All these entities are described by logical formulas.

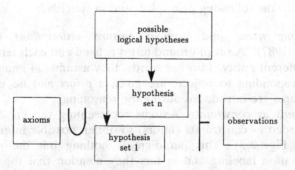

Figure 2.10. Logical framework for hypothesis-based reasoning.

Here we use the term *logical hypotheses* to discriminate them from our hypotheses.

Since the given knowledge is incomplete, the reasoner cannot derive (prove) the observations based on the axioms alone. Then, it searches for a set of logical hypotheses which satisfy the following conditions:

I. {axioms} ∪ {logical hypotheses} ⊢ {observations}

II. {axioms} ∪ {logical hypotheses} are logically consistent

Here A ⊢ B means that B can be derived from A by applying inference rules, and the logical consistency in (II) implies that no contradiction is derived from {axioms} ∪ {logical hypotheses}: both ¬P and P are not derived. (We assume {axioms} itself is logically consistent.) Note that logical consistency is different from the consistency examined by GRE in SIGMA; the consistency examined by GRE implies the equality relation between ground terms, while the logical consistency here implies that no contradiction is logically derived from a set of formulas.

The goal of hypothesis-based reasoning is to find a set of logical hypotheses which complement the incompleteness of the knowledge. That is, the reasoner constructs *explanations* of the observations in terms of both the given knowledge and the selected logical hypotheses. This reasoning scheme is very different from ordinary deductive reasoning, in which {axioms} ⊢ {theorems to be proved} is assumed. That is, since given {axioms} and inference rules, the set of derivable theorems is fixed, what is done in deductive reasoning is to examine if {derivable theorems} ∈ a theorem to be proved. In other words, ordinary deductive reasoning does not generate new information, while the task of hypothesis-based reasoning is to find new information (i.e., logical hypotheses) to explain the observations. As will be discussed later, since many sets of logical hypotheses can satisfy (I) and (II), as shown in Fig. 2.10, the reasoner in hypothesis-based reasoning has to select a *preferable* one from among them.

The entire reasoning process in SIGMA can be formulated based on this reasoning scheme (Fig. 2.11). In SIGMA, knowledge is classified into three types: knowledge about the scene, knowledge about the mapping between the scene and the image, and knowledge about the image. The observed image data in SIGMA correspond to the observations in Fig. 2.10. The logical hypotheses in SIGMA are described by two types of formulas: equality relations between ground terms like $b = f(a)$ and unary ground predicates like $P(a)$. The former type of logical hypotheses are generated by GRE, and the latter by MSE.

LLVE analyzes the image data to extract image features. An image

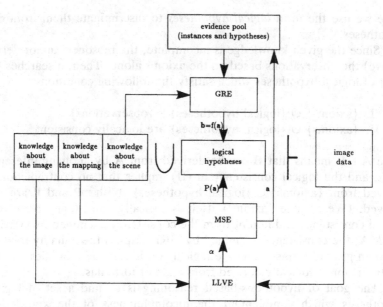

Figure 2.11. Logical framework of the reasoning in SIGMA.

feature in this logical framework is represented by a constant, say α. Note that constants representing image features should be clearly discriminated from those representing object instances. MSE transforms α to a constant representing an object instance, say **a** based on the knowledge about the mapping. In other words, we apply a transformation function *trans* to α to obtain $\mathbf{a}: \mathbf{a} = trans(\alpha)$. Then, MSE inserts constant **a** into the evidence pool and generates a logical hypothesis like $P(\mathbf{a})$.

The evidence pool, which corresponds to the Iconic/Symbolic Database, stores all ground terms generated so far: constants generated by MSE and ground functions instantiated by GRE. Logically speaking, the evidence stored in the evidence pool is a subset of the *Herbrand universe* (Mend1964, Gene1987). GRE examines the equality between ground terms in the evidence pool to generate a logical hypothesis like $\mathbf{b} = f(\mathbf{a})$. Based on these logical hypotheses and knowledge about the scene, ground relational predicates like $REL(\mathbf{a}, \mathbf{b})$ are derived. This derivation process corresponds to the construction of the scene description.

The constructed scene description can be back-projected to the image description which matches the observed image data. Knowledge about the mapping includes reverse functions of transformation functions like *trans*. Such reverse functions are used for the back-projection.

In the top-down analysis, MSE uses such reverse functions to transform descriptions about ground functions [i.e., hypotheses like $f(\mathbf{a})$] in the scene domain into those in the image domain. That is, a hypothesis in the scene domain $f(\mathbf{a})$ is transformed into that in the image domain $trans^{-1}(f(\mathbf{a}))$ by MSE. LLVE analyzes the image to extract the image feature β which satisfies $\beta = trans^{-1}(f(\mathbf{a}))$. Then, MSE creates a new constant $\mathbf{c}$ in the scene domain which satisfies $\mathbf{c} = trans(\beta)$, and GRE generates a new logical hypothesis $\mathbf{c} = f(\mathbf{a})$. Thus in the top-down analysis, descriptions are transformed from the scene domain to the image domain and vice versa, and new constants in the scene and image domains and logical hypotheses are created.

Note that geometric relations among image features are used only internally by LLVE, so that they are not transformed to those among scene objects. In Reit1987, on the other hand, geometric relations among image features are transformed to those among scene objects on knowledge about the mapping. For example, their knowledge includes the following axiom:

$$\forall x, y[\Delta(x, \sigma(x)) \wedge \Delta(y, \sigma(y)) \supset interior(x, y) \equiv \textit{INSIDE}(\sigma(x), \sigma(y))] \tag{2.35}$$

where variables x and y denote image features, $\sigma(x)$ and $\sigma(y)$ denote their corresponding object instances (i.e., predicate Δ specifies such correspondence between an image feature and an object instance), predicate *interior* specifies a geometric relation in the image domain, and *INSIDE* specifies its corresponding relation in the scene domain. $\equiv$ means logical *EQUIVALENCE*. With this axiom, given α and β (i.e., image features) satisfying $interior(\alpha, \beta)$, the theorem $\textit{INSIDE}(\sigma(\alpha), \sigma(\beta))$, which describes a geometric relation in the scene domain, is derived. In general, however, it is very difficult to symbolically describe the correspondence between geometric relations in the image domain and those in the scene domain. In particular, when the scene is 3D, such correspondence cannot be established; separate 3D objects can be projected as neighboring regions in the image, and connected 3D objects can be separated in the image due to occlusions. This is the reason why we do not use any knowledge about the mapping of geometric relations, although the current implementation of SIGMA analyzes 2D scenes.

The discussion in the preceding paragraph implies an important characteristic of hypotheses in SIGMA. As discussed in Section 2.3.1, a set of constraints are generated for each hypothesis [e.g., $f(\mathbf{a})$] based on axioms like (2.10). In SIGMA all such constraints are on intrinsic properties (i.e., attributes) of target objects. No constraints on geometric relations to other objects are associated with hypotheses; SIGMA cannot transform geometric relations from the scene domain to the image domain and vice versa. For example, suppose a house instance **a** generates a hypothesis about *DRIVEWAY*, $f(\mathbf{a})$. **a** computes constraints on attributes of the target driveway, such as location, width, and length, which are associated with $f(\mathbf{a})$. Although there exists the following geometric constraint on the target driveway, **a** does not associate it with $f(\mathbf{a})$: the target driveway must be connected to **a**. **a** examines such geometric constraints when an instance satisfying $f(\mathbf{a})$ is returned from GRE. That is, let **b** denote a driveway instance satisfying $f(\mathbf{a})$ [i.e., $\mathbf{b} = f(\mathbf{a})$]. **a** examines whether **b** is connected to itself. If so, **a** establishes a spatial relation with **b**. Otherwise, **a** initiates a new reasoning based on the fact that although a possible candidate for its driveway has been detected, the candidate is not connected to **a**. Section 3.9 describes a detailed reasoning process to manage geometric constraints.

2.3.6. Discussion

In the logical framework described in the previous section, we should note the following points:

1. *Examination of logical consistency.* In hypothesis-based reasoning, we have to find a set of logical hypotheses satisfying the two conditions (I) and (II) described before. From a theoretical point of view, however, it is not possible to verify the logical consistency among an arbitrary set of logical formulas in the first-order predicate calculus. That is, we have no general algorithm to verify condition (II).

From a practical point of view, on the other hand, we know various causes of contradiction and constraints not to be violated in a specific application domain. Thus, using such knowledge, we can implement domain-specific procedures to examine the logical consistency. That is, we consider {axioms} ∪ {logical hypotheses} as logically consistent unless it violates the domain-specific constraints.

In SIGMA, we use the knowledge that different object instances cannot be located at the same location as the constraint to examine the logical consistency. We will discuss this point in Section 2.4.4.

As will be described in Section 3.4, since GRE has to generate logical hypotheses which cause no contradiction (i.e., prevent mutually conflicting pieces of evidence from being accumulated), its consistency examination process becomes rather complicated.

2. *Multiple possible explanations.* As illustrated in Fig. 2.10, there can be many sets of logical hypotheses which explain the same observations. In SIGMA, multiple descriptions of the scene are constructed corresponding to these hypothesis sets.

One idea to reduce the number of hypothesis sets is to get new observations by performing new measurements and/or asking a user about the world; hypothesis sets which cannot explain the new observations are discarded. In image understanding, however, it is very difficult to obtain complete information about the scene, as discussed in Section 1.3 (e.g., occlusions are inevitable). This implies that it is essential for image understanding systems to construct multiple scene descriptions.

Considering various *ambiguous figures* such as the Necker cube and Boring's mother-in-law (Greg1970) (i.e., figures which are perceived in multiple ways), the human visual system also seems to construct multiple interpretations and switch from one interpretation to another from time to time. The multiple scene descriptions that are constructed correspond to such competing interpretations, which cannot coexist at the same time. In other words, sets of logical hypotheses based on which computing descriptions are derived are mutually contradicting.

Poole proposed a criterion to select the most preferable hypothesis set (Pool1985). He compared the *specificity* of hypothesis sets and selected the most specific one as the preferable explanation. As will be discussed in Section 3.1.4, in SIGMA we use the *size* of constructed scene descriptions as the criterion to select the most preferable one. However, it should be noted that competing interpretations as discussed above have the same size according to our criterion, so that there may be still multiple descriptions left even after the selection. We should accept all such descriptions as final analysis results.

3. *Erroneous observations.* As discussed in Section 1.3, one of the difficult problems in image understanding is that erroneous image features are extracted due to noise. Here we discuss their effects on the logical framework of the hypothesis-based reasoning.

In Fig. 2.11 we included the segmentation process by LLVE in the framework so as to describe the entire reasoning process in SIGMA. Here in order to clarify the effects of erroneous image features, we exclude LLVE and use Fig. 2.12 as the logical framework: it illustrates the logical framework for the reasoning in the scene domain. That is, we confine ourselves to spatial reasoning and object recognition, excluding segmentation from the framework. Thus, the image data in Fig. 2.11 are replaced by image features in Fig. 2.12, which correspond to the observations to be explained.

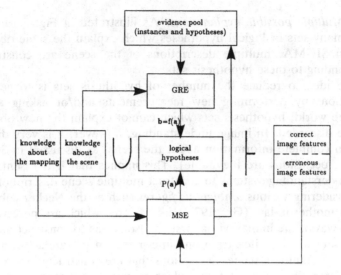

Figure 2.12. Reasoning with erroneous observations.

The framework in Fig. 2.10 assumes that all observations are valid and must be explained based on axioms and logical hypotheses. In Fig. 2.12, however, the observations (i.e., image features) usually include erroneous ones, so that the original framework of hypothesis-based reasoning must be modified to cope with such erroneous observations.

a. Relaxing condition (I). First we should relax condition (I) as follows:

(I)′ {axioms} ∪ {logical hypotheses} ⊢ a subset of {observations}

This is because, since the observations may include erroneous image features, not all observations need to be explained: since erroneous image features are meaningless, no interpretations for them are included in the *correct* scene description. In other words, we should accept a set of logical hypotheses even if it can explain only a subset of the observations.

In this modified framework, to select a certain set of logical hypotheses means that we regard as correct image features the set of observations which can be explained by using the selected logical hypotheses, and the others as erroneous ones. Thus, the selection of hypothesis sets becomes crucial; it should be regarded as the process of discriminating correct observations from erroneous ones.

b. Introduction of a new condition. Relaxing condition (I) causes an explosion of hypothesis sets. Let $\{\alpha_1, \alpha_2, \ldots, \alpha_n\}$ denote the entire set of observations. Then, in the worst case, 2^n hypothesis sets will be generated if all observations are correct; we will have a hypothesis set for each subset of $\{\alpha_1, \alpha_2, \ldots, \alpha_n\}$. Thus we need to introduce another condition to avoid such an explosion of hypothesis sets. (Recall that the goal of the original hypothesis-based reasoning is to find hypothesis sets which explain all the observations.)

In SIGMA, we introduce the concept of the *maximal set of explainable observations* and modify the goal of the reasoning as follows:

Find a set of logical hypotheses which can explain the largest subset of the observations.

We call such a subset the maximal set of explainable observations. In other words, we introduce an extralogical criterion to select the most preferable set of logical hypotheses. In a sense, this criterion measures the *size* of the constructed scene description; the more image features can be explained, the more logical hypotheses are generated and consequently the larger is the scene description constructed. Note again that all observations which are not included in the maximal set of explainable observations are regarded as erroneous ones.

4. *Negative and disjunctive knowledge.* As readers familiar with logic will have noticed, all axioms used in this section can be described by Horn clauses (Cloc1981, Lloy1984). Syntactically, a Horn clause is a clause with at most one positive literal (i.e., predicate) and it can be described as

$$P_1 \wedge P_2 \wedge \ldots P_n \supset P, \tag{2.36a}$$

$$P_1 \wedge P_2 \wedge \ldots P_n \supset , \tag{2.36b}$$

or

$$P \tag{2.36c}$$

where P and P_i are predicates and their arguments are omitted. Pure Prolog (Cloc1981) programs are typical examples of Horn clauses and are useful to represent various knowledge. However, they cannot represent negative or disjunctive knowledge. For example,

$$\forall x \{HUMAN(x) \wedge \neg MALE(x) \supset FEMALE(x)\} \tag{2.37}$$

and

$$\forall x \{HUMAN(x) \supset MALE(x) \vee FEMALE(x)\} \tag{2.38}$$

are not Horn clauses; each axiom is transformed to the same clausal form with two positive laterals.

Thus, the axioms we have used so far do not include any negative or disjunctive knowledge, but we need such knowledge to represent the world model. For example, a *HOUSE* has its facing *ROAD* either at its front or at its back, and a *ROAD* ends either at an intersection or at a dead end.

In SIGMA, such disjunctive knowledge is used internally by object instances: the overall spatial reasoning performed by object instances and GRE uses Horn clauses alone. Usually, each object instance selects one possible case from those specified by the disjunctive knowledge, and generates a hypothesis based on the selected case. If the hypothesis is verified by GRE, the instance accepts it and discards the other possibilities. Otherwise, it retracts the hypothesis and generates a new hypothesis by selecting another possible case based on the knowledge. Thus, disjunctive knowledge is used in *failure-driven reasoning,* in which the failure of a certain reasoning process triggers another reasoning process based on disjunctive knowledge. In this sense, the disjunction is regarded as an exclusive *OR*. Detailed descriptions of how to represent disjunctive knowledge in SIGMA will be given in Section 3.8.

As for negative knowledge, SIGMA does not use it at all; all pieces of evidence accumulated are positive information and no negative evidence is generated. This is because, while negative knowledge plays an important role in such applications as medical diagnosis (e.g., the presence of some symptoms precludes the possibility of certain diseases), we could not see any typical examples of such useful negative knowledge in image undestanding.

2.4. OBJECT-ORIENTED KNOWLEDGE REPRESENTATION

As we have discussed, many types of knowledge are required for IUSs to interpret images. In SIGMA, the following types of knowledge are used:

1. Knowledge about attributes of objects, e.g., location, area size, and shape features of a house.
2. Knowledge about relations between objects, e.g., driveways connect roads and houses, and a house-group consists of a set of regularly arranged houses.
3. Control knowledge to guide the analysis, e.g., when the front of a house is determined, check the relation between the house and its driveway.

Our knowledge representation scheme has its root in the frame system (Mins1975) and the object-oriented computation paradigm (Wein1980, Gold1983). In SIGMA, models of objects in the world are described by *object classes*. An object class consists of *slots, links,* and *rules.* Slots are used to record values of object properties. Hierarchical relations between objects in the world are described using links between classes. Spatial relations between objects and the control knowledge are described by rules.

In this section, we describe each of these representational primitives.

2.4.1. Object Classes and Slots

The basic entities of knowledge representation in SIGMA are called *object classes*. They model abstract objects in the problem domain such as *HOUSE* and *ROAD*. When an object in the scene is recognized, the corresponding object class is *instantiated* to produce an *object instance* representing the recognized object. That is, an object class describes an abstract model of a class of objects in the world, while an object instance represents an object observed in the scene (i.e., recognized from the image).

An object class not only describes properties and structures of objects in the world but also stores the knowledge to conduct object recognition, i.e., recognition of related objects. Such knowledge is described by rules. An object instance performs reasoning by using rules stored in its corresponding object class. Thus, in our scheme, each object instance is not a static data structure to record properties of a recognized object but an active reasoning agent which performs reasoning based on its own knowledge. This is why we call our knowledge representation scheme object-oriented.

Each object class has three fields: *slots, links,* and *rules.* Slots are used to record property values of a recognized object. Hierarchical relations between objects in the world are described using links between object classes. Spatial relations between objects and the control knowledge are described by rules.

Each object class may have many associated descriptions that define semantic properties of the object in the world: attributes and symbolic relations to other objects. These properties are described by *slots*. Slots are similar to property lists in Lisp. Each slot is a list which contains an indicator (i.e., slot name) and a value.

For example, the following object class definition

> object-class RECTANGULAR-HOUSE;
>   slots: centroid;
>   shape-description;
>   front-of-house;
>   front-road;
>   connecting-driveway;
>   end-object-class

defines an object class named RECTANGULAR-HOUSE, which includes five slots: centroid, shape-description, front-of-house, front-road, and connecting-driveway. Note that the former three slots describe attributes and the latter two symbolic relations to other objects: for example, the front-road slot is used to store the symbolic relation to (i.e., ID number of) the front road instance.

2.4.2. Rules for Spatial Reasoning

In addition to slots in which property values are recorded, we associate with each object class knowledge about how to compute values of the slots. We represent this type of knowledge by *rules*. In SIGMA, most of the domain knowledge required to understand the image is represented by rules.

As discussed in Section 2.2.2, a spatial relation between objects is represented by a pair of functions. The most common use of rules is to represent such functions together with the control information about when to apply the functions. In other words, in describing rules, we first compile spatial relations in the world into functions, and then associate the control knowledge with them. When activated, a rule generates a hypothesis about a related object, and then fills the corresponding slot (e.g., the front-road slot in the above example) with the ID number of the object instance which satisfies the hypothesis.

In our scheme, a rule is composed of three parts:

> ⟨Control-condition⟩
>
> ⟨Hypothesis⟩
>
> ⟨Action⟩

⟨Control-condition⟩ is a predicate. It indicates when the rule can potentially be applied. ⟨Hypothesis⟩ specifies a procedure to generate an expected description (i.e., hypothesis) of a related object. That is, functions described in Setion 2.2.2 to generate hypotheses are stored in ⟨hypothesis⟩ parts of rules. ⟨Action⟩ describes a procedure to be executed when the hypothesis generated by the ⟨hypothesis⟩ part is verified/refuted.

In the interpretation process, first the ⟨control-condition⟩ of each rule is examined and, if it evaluates true, the procedure in the ⟨hypothesis⟩ part is executed to generate and insert a hypothesis into the Iconic/Symbolic Database. Then, a solution for the hypothesis is computed by GRE. Finally, the procedure in the ⟨action⟩ part is executed using the computed solution. In general, ⟨action⟩ parts can add facts to or delete facts from the Iconic/Symbolic Database.

For example, suppose we have the following house–road relation:

A road $road_x$ is *along* a house $house_x$ if the predicate $along(road_x, house_x)$ is true.

Here $road_x$ and $house_x$ denote arbitrary object instances.

This piece of knowledge can be encoded in the following rule (say $R_{house-road}$), which is stored in the object class *HOUSE*:

$R_{house-road}$:

⟨Control-condition⟩: *true*

⟨Hypothesis⟩: $P_{house-to-road}$

⟨Action⟩: *self.front-road = solution*

Note that the similar rule (say $R_{road-house}$) is stored in *ROAD*, by which an instance of *ROAD* generates a hypothesis for *HOUSE*.

Suppose we have a house instance $house_0$. $house_0$ performs reasoning using $R_{house-road}$ to detect the road instance which satisfies the relation *along* with itself. The rule $R_{house-road}$ is applied whenever *HOUSE* is instantiated, since the ⟨control-condition⟩ part is set to *true*. $P_{house-to-road}$ represents a procedure which generates a hypothesis for an unknown road instance $road_x$. The procedure uses the properties of $house_0$ to generate the description of the hypothesis. The hypothesis is described in a similar way as object instances:

$road_x$: hypothesis for *ROAD*
target-object-class
ROAD

orientation
    greater than $house_0$.front-of-house $+ 80°$
    and less than $house_0$.front-of-house $+ 90°$
width
    greater than $house_0$.width $\times 0.3$
    and less than $house_0$.width $\times 0.5$
centroid
    resides within $REGION(house_0$.centroid, $house_0$.front-of-house)

Here $house_0$.front-of-house, $house_0$.width, and $house_0$.centroid denote corresponding slot values of $house_0$. $REGION$ is a function to generate the expected location of the target road (Fig. 2.13).

In $R_{house\text{-}road}$, *solution* denotes the solution computed by GRE for the hypothesis $road_x$. When $R_{house\text{-}road}$ is used by $house_0$, variable *self* is bound to $house_0$. The ⟨action⟩ part of $R_{house\text{-}road}$ indicates that the front-road slot should be filled with *solution* once it is computed.

Suppose we want to encode another piece of knowledge about the house–road–driveway relation:

A driveway $driveway_x$ *connects* house $house_x$ and road $road_x$ if $along(road_x, house_x)$ is true and the predicate $connect(driveway_x, road_x, house_x)$ is true.

This piece of knowledge can be encoded in the following rule (say $R_{house\text{-}road\text{-}driveway}$), which is also stored in the object class $HOUSE$:

    $R_{house\text{-}road\text{-}driveway}$

    ⟨Control-condition⟩: $along(self.front\text{-}road, self)$

    ⟨Hypothesis⟩: $P_{connect\text{-}house\text{-}road}$

    ⟨Action⟩: $self.connecting\text{-}driveway = solution$

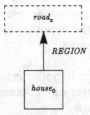

Figure 2.13. Generation of the expected location of a road.

The ⟨control-condition⟩ part indicates that the rule is to be applied when the front-road slot of a house instance is filled and the predicate *along(self.front-road, self)* is true. That is, this rule is not activated until a front road instance of a house instance is detected. $P_{connect-house-road}$ is a procedure which generates the following hypothesis $driveway_x$:

> $driveway_x$: hypothesis for DRIVEWAY
>> target-object-class
>>> DRIVEWAY
>> orientation
>>> greater than $house_0$.front-of-house $- 10°$
>>> and less than $house_0$.front-of-house $+ 10°$
>> width
>>> greater than $house_0$.width $\times 0.3$
>>> and less than $house_0$.width $\times 0.5$
>> centroid
>>> resides with CONNECT-REGION(*self.front-road, self*)

CONNECT-REGION computes the expected location of $driveway_x$ between the road and the house (Fig. 2.14). The ⟨action⟩ part of $R_{house-road-driveway}$ indicates that the solution is to be filed in the connecting-driveway slot.

Let $R_{house-road}$ and $R_{house-road-driveway}$ be the rules described in the above. The object class RECTANGULAR-HOUSE can now be described as follows (the relation between HOUSE and RECTANGULAR-HOUSE will be discussed in the next section):

> *object-class* RECTANGULAR-HOUSE;
>> *slots*: centroid;
>>> shape-description;
>>> front-of-house;
>>> front-road;
>>> connecting-driveway;
>> *rules*: $R_{house-road}$;
>>> $R_{house-road-driveway}$;
> *end-object-class*

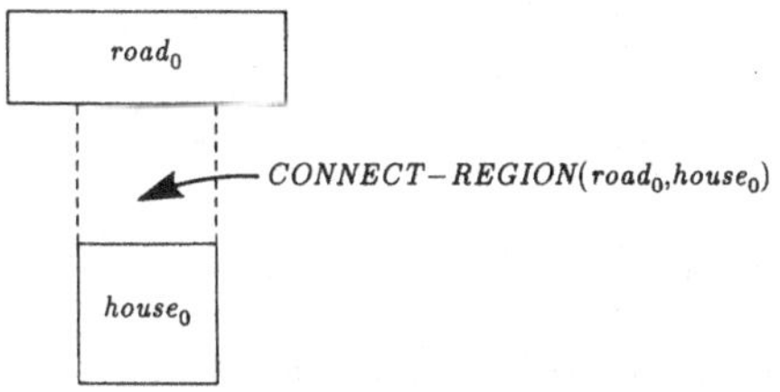

Figure 2.14. Generation of the expected location of a driveway.

Note that the rules described above represent unary or binary relations (i.e., a relation between no more than two objects). To describe an *n*-ary relation (i.e., a relation among *n* objects), we need to translate it into a set of binary relations. One convenient way to do the translation is to introduce an imaginary object X and describe the *n*-ary relation by a set of binary relations between X and each object:

$$\text{REL}(O_1, O_2, \ldots, O_n) \Rightarrow \text{REL}_1(X, O_1), \text{REL}_2(X, O_2), \ldots, \text{REL}_n(X, O_n)$$

As described in Section 2.2.3, similar spatial reasoning is performed based on *PART-OF* relations. It differs from that based on spatial relations in several ways: (1) bottom-up instantiation of a parent object, (2) top-down hypothesis generation, and (3) unification of partially instantiated *PART-OF* hierarchies. The knowledge required for these reasoning processes is also described by the same type of rules as are used for reasoning based on spatial relations.

Let O_{whole}, O_{part}, and $O_{subpart}$ denote object classes representing a whole object, its part, and its subpart in a *PART-OF* hierarchy, respectively (Fig. 2.15). When an instance of $O_{subpart}$, say $I_{subpart}$, is generated, it applies a rule stored in $O_{subpart}$ to directly generate its parent instance of O_{part}, say I_{part}. The ⟨hypothesis⟩ part of this rule is set NIL and the ⟨action⟩ part generates the parent instance. Thus, the rule directly instantiates O_{part} without generating a hypothesis. The bottom-up instantiation is realized by using such rules.

As described in Section 2.2.3, newly generated instance I_{part} is either *partially instantiated* or *fully instantiated*. The state of I_{part} is determined by comparing the kernel list stored in O_{part} with its constituent object instances (i.e., $I_{subpart}$ in this example). If I_{part} is fully instantiated, it also

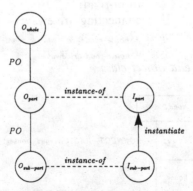

Figure 2.15. A *PART-OF* hierarchy.

generates its parent instance of O_{whole}. A rule stored in O_{part} is used to perform these reasoning processes: the determination of the state of an instance and the instantiation of a parent object. Specifically, the ⟨control-condition⟩ part of the rule examines the state of I_{part} based on the kernel list in O_{part}, the ⟨hypothesis⟩ part is NIL, and the ⟨action⟩ part generates a parent object instance of O_{wide}.

I_{part} can apply another rule in O_{part} to generate a hypothesis for a missing subpart. This top-down hypothesis generation is exactly the same as the hypothesis generation based on spatial relations.

As illustrated in Fig. 2.7, a part object instance is sometimes connected to multiple parent instances by PART-OF relations. In such a case, the part instance examines the consistency between the parent instances, and if they are consistent, they are unified into one instance. The knowledge required for this reasoning process is also described by a rule, which is stored in the part object class. The part instance uses the rule to realize the unification when it finds that it has multiple parent instances: the ⟨control-condition⟩ of the rule examines if there are multiple parent instances connected by PART-OF relations. This type of rule is different from those used for ordinary spatial reasoning. Since the detailed control mechanism of the system must be explained to understand how such rules are used, we will describe the unification process in detail in Section 3.6.4.

In summary, various types of rules are stored in object classes, based on which their instances perform reasoning: hypothesis generation and establishment of symbolic relations based on spatial relations, bottom-up instantiation and top-down hypothesis generation based on PART-OF relations, and the unification of multiple parent instances. There are still other types of rules used for object instances to perform reasoning. Section 3.7 describes a taxonomy of rules with various illustrative examples.

2.4.3. Links for Organizing Knowledge

Objects in the world are often organized into hierarchies. It is natural and convenient to preserve these hieararchies when we construct the world model. In our knowledge representation, *links* are used to organize object classes into hierarchical structures.

In SIGMA, three types of links are used to represent semantic relations between object classes: A-KIND-OF, APPEARANCE-OF, and PART-OF links. A-KIND-OF and PART-OF links connect pairs of object classes to form the hierarchical world model. An APPEARANCE-OF link connects an object class in the scene domain with that in the image domain representing its appearance. Links are used mainly by MSE to perform reasoning for appearance model selection.

Although we have been using the term *relations* intuitively so far, we will differentiate links from relations as follows. Links connect objects classes, while relations connect object instances. For example, *HOUSE* is connected with *RECTANGULAR-HOUSE* by an *A-KIND-OF* link, which describes the structure of the world model. On the other hand, a house instance is connected with a rectangular-house instance by an *A-KIND-OF* relation, which describes the structure of a constructed interpretation. As described below, the former are used to translate a description of an object class into that of another class, while the latter represent that object instances connected by them are satisfying specific geometric and semantic relations.

Attached to links are functions which translate a description of one object class into that of the other. That is, such functions define semantics of links. Since the translation is usually bidirectional, a pair of functions is associated with each link. In other words, we give different meanings to a link depending on the direction in which the link is used.

1. *A-KIND-OF Links.* One popular object hierarchy is the generalization/specialization hierarchy. This hierarchy is organized by connecting object classes by *A-KIND-OF* (*AKO*) links. Link $AKO_{specialization}$ connects an object class to its specialization while link $AKO_{generalization}$ connects an object class to its generalization (Fig. 2.16).

Suppose we have defined *RECTANGULAR-HOUSE* as described before, and want to define a new class of houses named *L-SHAPED-HOUSE*. This object class contains the same slots and rules as *RECTANGULAR-HOUSE* except the knowledge about shape descriptions: say, rule $R_{rectangle}$ computes the shape of *RECTANGULAR-HOUSE* and rule $R_{L\text{-}shape}$ computes the shape of *RECTANGULAR-HOUSE* and rule $R_{L\text{-}shape}$ computes the shape of

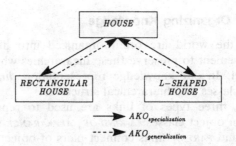

Figure 2.16. *A-KIND-OF* links between *HOUSE*, *RECTANGULAR-HOUSE*, and *L-SHAPED-HOUSE*.

L-SHAPED-HOUSE. Note that rules can be used to compute attributes of objects as well as to perform spatial reasoning as described in the previous section.

Instead of defining two almost identical object classes individually, we introduce a new abstract object class named *HOUSE* and use *AKO* links as follows:

> *object-class* RECTANGULAR-HOUSE;
>   *slots*: shape-description(rectangle);
>   *rules*: $R_{rectangle}$;
>   *links*: $AKO_{generalization}$: HOUSE;
> *end-object-class*

> *object-class* L-SHAPED-HOUSE;
>   slots: shape-description(L-shape);
>   *rules*: $R_{L-shape}$;
>   *links*: $AKO_{generalization}$: HOUSE;
> *end-object-class*

> *object-class* HOUSE;
>   slots: centroid;
>   front-of-house;
>   front-road;
>   connecting-driveway;
>   *rules*: $R_{house-road}$;
> $R_{house-road-driveway}$;
>   *links*: $AKO_{specialization}$: (RECTANGULAR-HOUSE, L-SHAPED-HOUSE);
> *end-object-class*

All the properties and rules defined in *HOUSE* are *inherited* to *RECTANGULAR-HOUSE* and *L-SHAPED-HOUSE* through $AKO_{generalization}$ links. That is, both *RECTANGULAR-HOUSE* and *L-SHAPED-HOUSE* have centroid, front-of-house, front-road, and connecting-driveway slots as well as their own shape-descriptions. Also, both of these object classes can use the rules $R_{house-road}$ and $R_{house-road-driveway}$ defined in *HOUSE*. Thus, $AKO_{generalization}$ links are used to efficiently encode the world knowledge. This inheritance mechanism is a major characteristic of object-oriented programming languages (Wein1980, Gold1983).

Hypotheses generated by rules are often described in terms of abstract object classes like *HOUSE*; for example, a road instance, which is going to generate a hypothesis for its facing house, usually does not have any information about a specific type of house to be expected, so that it generates a hypothesis about *HOUSE*. Therefore, a target object in the top-down request, which is passed to MSE from GRE, is also described by an abstract object class (e.g., *HOUSE*). In such case, MSE needs to perform reasoning using $AKO_{specialization}$ links in order to select an appropriate specific object class; although MSE needs to determine a concrete appearance of a target object, an abstract object class specified in a request has no shape information. For example, suppose GRE asks MSE to verify the existence of a house in a certain local area. MSE must determine the concrete shape of the target house before issuing the request to LLVE.

This appearance model selection can be achieved by following $AKO_{specialization}$ links from *HOUSE*. That is, MSE examines the $AKO_{specialization}$ link field of *HOUSE* and selects either *RECTANGULAR-HOUSE* or *L-SHAPED-HOUSE*. The search through $AKO_{specialization}$ links continues until a specific object class with a concrete shape description is found. (As will be described below, object classes with *APPEARANCE-OF* links are regarded as those with concrete shapes.)

In general, however, there are many possible choices in the search for a specialized object class (paths in the search through $AKO_{specialization}$ links). For example, MSE has to decide which of *RECTANGULAR-HOUSE* or *L-SHAPED-HOUSE* should be selected to specialize *HOUSE*. MSE needs the knowledge to determine the appropriate order in which the alternatives should be investigated. The knowledge to guide the search is called *specialization strategy*. It is encoded by an ordered list containing a set of all specialized object classes connected by $AKO_{specialization}$ links. The list is stored in an $AKO_{specialization}$ link field as shown in the preceding example of *HOUSE*. The list specifies the order in which MSE conducts the search. In the case of *HOUSE*, MSE first selects *RECTANGULAR-HOUSE* and searches for its 2D appearance. And if the top-down analysis based on the selected appearance fails, MSE tries *L-SHAPE-HOUSE* to find another possible appearance.

Attached to each $AKO_{specialization}$ link is a function, $F_{specialization}$, which translates a description of a general object class into that of its specialized object class. During the search for an object appearance in the top-down analysis, MSE applies such functions to translate a hypothesis of a general object class into that of a specialized object class. Then MSE passes a translated hypothesis to LLVE as its goal specification.

Now *HOUSE* is defined as:

> *object-class HOUSE*;
>   *slots*: centroid;
>   front-of-house;
>   front-road;
>   connecting-driveway;

$rules$: $R_{house\text{-}road}$

   $R_{house\text{-}road\text{-}driveway}$;

$links$: $AKO_{specialization}$: $(RECTANGULAR\text{-}HOUSE, L\text{-}SHAPED\text{-}HOUSE)$;

   $F_{specialization}$: $(F_{RECTANGULAR\text{-}HOUSE}, F_{L\text{-}SHAPED\text{-}HOUSE})$;

$end\text{-}object\text{-}class$

GRE examines the consistency between pieces of evidence in the Iconic/Symbolic Database. As will be described in Section 3.4.2, object classes of some pieces of evidence are at different levels in an AKO hierarchy. In order to examine the consistency between such pieces of evidence, GRE translates a description of a specialized object class into that of its generalized object class. In order to realize this translation, a function $F_{generalization}$ is attached to each $AKO_{generalization}$ link. Thus $RECTANGULAR\text{-}HOUSE$ is defined as follows:

$object\text{-}class$ $RECTANGULAR\text{-}HOUSE$;

   $slots$: shape-description(rectangle);

   $rules$: $R_{rectangle}$;

   $links$: $AKO_{generalization}$; $HOUSE$;

      $F_{generalization}$: F_{HOUSE};

$end\text{-}object\text{-}class$

2. $APPEARANCE\text{-}OF$ $Links$. In SIGMA, GRE performs reasoning in the scene domain while LLVE performs reasoning in the image domain. That is, all the object classes in the world model are described in terms of scene domain terminology, while the knowledge used by LLVE is described in terms of image domain terminology. MSE bridges these two levels of reasoning and translates the terminology in the scene domain into that in the image domain and vice versa. Thus we need the knowledge for this translation.

In our knowledge representation scheme, $APPEARANCE\text{-}OF$ links represent rules for the translation: object classes in the scene domain are connected via $APPEARANCE\text{-}OF$ links to those in the image domain, which represent appearances of objects in the image. For example, let $RECTANGLE$ be an object class in the image domain that represents rectangularly shaped regions. Since, viewed from above, the appearance of $RECTANGULAR\text{-}HOUSE$ can be described by a rectangularly shaped region, we establish an $APPEARANCE\text{-}OF$ link between $RECTANGULAR\text{-}HOUSE$ and $RECTANGLE$. That is, we introduce an $APPEARANCE\text{-}OF$ link in object class $RECTANGULAR\text{-}HOUSE$ as follows:

$object\text{-}class$ $RECTANGULAR\text{-}HOUSE$;

   $slots$: shaped-description(rectangle);

   $rules$: $R_{rectangle}$;

$links$: $AKO_{generalization}$: $HOUSE$;
$F_{generalization}$: F_{HOUSE};
$APPEARANCE\text{-}OF$: $RECTANGLE$;
$F_{first\text{-}order\text{-}properties}$: $F_{HOUSE\text{-}RECTANGLE}$;
$end\text{-}object\text{-}class$

Attached to each $APPEARANCE\text{-}OF$ link is a function that translates a description of the object class in the scene domain into that of the object class in the image domain. Let $F_{first\text{-}order\text{-}properties}$ denote such a function. The function is stored in the object class in the scene domain as illustrated previously, and models the geometric transformation done by the camera.

During the search through $AKO_{specialization}$ links (i.e., the search for an appropriate specialized object class in the top-down analysis), MSE selects the first encountered object class with an $APPEARANCE\text{-}OF$ link as the best candidate to try. Then, MSE applies $F_{first\text{-}order\text{-}properties}$ attached to the $APPEARANCE\text{-}OF$ link to translate the hypothesis in the scene domain into that in the image domain. For example, when MSE is given a $RECTANGULAR\text{-}HOUSE$ hypothesis, it applies $F_{HOUSE\text{-}RECTANGLE}$ stored in $RECTANGULAR\text{-}HOUSE$ to generate an appropriate $RECTANGLE$ hypothesis that describes its appearance in the image.

In general, since an object appearance can change depending on photographic conditions, such as viewing angle, focal length, or resolution of the camera, an object class in the scene domain is connected to many different object classes in the image domain by $APPEARANCE\text{-}OF$ links. Thus, MSE needs to determine which object class in the image domain is the most plausible appearance of the target object. Although this reasoning is crucial in recognizing 3D objects, we did not implement it in the current version of SIGMA; it was implemented for 2D aerial image understanding. That is, an object class in the scene domain has at most one $APPEARANCE\text{-}OF$ link.

As Minsky (Mins1975) proposed, to represent a 3D object by a group of its 2D appearances is a useful modeling method in image understanding. Recently, Ikeuchi (Ikeu1987) applied such an appearance-based model in his object recognition system. In this sense, it is not hard to augment SIGMA for 3D object recognition: associate a set of all possible appearances with each object class via $APPEARANCE\text{-}OF$ links, and incorporate into MSE the reasoning process of selecting an appropriate appearance.

3. $PART\text{-}OF$ Links. When an object to be modeled has complex internal structure, it is often convenient to describe the object by

decomposing it into a group of simple parts. *PART-OF* (*PO*) links are used to specify such structural composition. Like *AKO* links, *PO* links are used to translate a description of a complex object class into that of its part object class and vice versa.

Link $PO_{whole-part}$ indicates how a complex object (i.e., whole) can be decomposed into a group of simpler objects (i.e., parts). That is, just as $AKO_{specialization}$ links, a whole object class contains a $PO_{whole-part}$ link field, where a list of its part object classes are recorded. Attached to the list of $PO_{whole-part}$ links are a group of functions ($F_{whole-part}$s) to compute descriptions of individual part objects based on that of the whole object.

In appearance model selection by MSE $F_{whole-part}$s are used to translate a hypothesis about a whole object class into hypotheses about its part object classes. In spatial reasoning by GRE, on the other hand, they are used to generate hypotheses for part objects based on an instance of a whole object class. As discussed in the previous section, when an instance of a whole object class is generated, it applies a rule to generate a hypothesis for its part object. Function $F_{whole-part}$ is used in the ⟨hypothesis⟩ part of such rule to generate a hypothesis.

For example, suppose that *CAR* consists of *BODY*, *TIRE*, and *DOOR* (Fig. 2.17). Functions $F_{CAR-BODY}$, $F_{CAR-TIRE}$, and $F_{CAR-DOOR}$ transform a description of *CAR* into respective descriptions of the part objects. Suppose that MSE is given a hypothesis about *CAR*. MSE uses the three functions to decompose the original complex hypothesis into three simpler hypotheses about *BODY*, *TIRE*, and *DOOR*. On the other hand, suppose an instance of *CAR* is generated by an instance of *BODY* (see Fig. 2.6). The *CAR* instance applies $F_{CAR-TIRE}$ and $F_{CAR-DOOR}$ to itself to generate hypotheses about the missing parts (i.e., *TIRE* and *DOOR*).

Link $PO_{part-whole}$ indicates how a part object can construct the description of its whole object. Attached to $PO_{part-whole}$ is a function ($F_{part-whole}$) which specifies how such construction should be performed (i.e., how to translate a description of a part object class into that of its

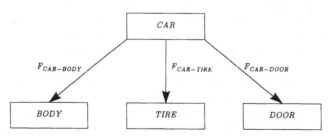

Figure 2.17. *PART-OF* links between *CAR* and its part objects.

whole object class). That is, each part object class includes a $PO_{part-whole}$ link field, in which the name of its whole object class is recorded. When an instance of the part object is generated, $F_{part-whole}$ is used to instantiate the whole object. Specifically, a part instance applies to itself a rule to generate a whole instance to which it belongs. $F_{part-whole}$ is used in the ⟨action⟩ part of such rule. (See the discussion in the previous section for how such a rule is activated.)

Our knowledge representation scheme contains two different ways of performing spatial reasoning: one based on spatial relations and the other based on PO relations. Although the knowledge used for both types of spatial reasoning is represented by rules, rules for spatial reasoning based on PO relations are closely related to PO links between object classes. In other words, we can consider that such rules are associated with P links and define the semantics of the links.

We usually illustrate the world model by a network, where nodes represent object classes and arcs various relations: PART-OF, A-KIND-OF, APPEARANCE-OF, and spatial relations. In such a network representation, an arc representing a PART-OF relation means a PART-OF link associated with corresponding rules. Arcs for A-KIND-OF and APPEARANCE-OF relations mean A-KIND-OF and APPEARANCE-OF links associated with corresponding translation functions. Although there are no links corresponding to spatial relations, we connect a pair of object classes (i.e., nodes) by an arc when rules for spatial reasoning based on spatial relations are stored in the object classes.

Since hypotheses are generated independently based on both spatial relations and PO relations, there are several restrictions on the knowledge organization. First, a PO hierarchy must be a tree structure. Second, spatial relations represented by rules must not be established between object classes in the same PO hierarchy. In other words, spatial relations must be established across different PO hierarchies. For example, consider the knowledge organization in Fig. 2.18a. There are two paths for object class A to generate a hypothesis of object class B: one by the spatial relation and the other by PO relations. This means that when an instance of A is generated, two different hypotheses for B can be generated from the same instance. If we allowed such hypothesis generation, multiple pieces of evidence generated from the same source would be accumulated. Third, multiple spatial relations must not be established across a pair of PO hierarchies (Fig. 2.18b). The same argument as in Fig. 2.18a holds. Note that a circular path consisting of spatial relations alone is allowed (Fig. 2.18c), since no redundant hypotheses are generated.

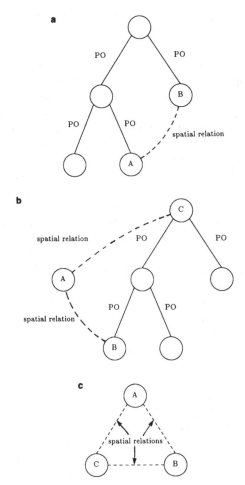

Figure 2.18. Avoiding generation of redundant hypotheses. (a) Spatial relation between object classes in the same *PO* hierarchy. (b) Multiple spatial relations to object classes in the same *PO* hierarchy. (c) Circular path consisting of spatial relations alone.

2.4.4. Mutually Conflicting Objects

Some pairs of objects cannot occupy the same location in the scene. For example, a region in an image cannot be interpreted as both a house and a road at the same time. On the other hand, shadows can be cast on any type of object. Moreover, when a road and a river intersect, we can recognize the intersecting area as a bridge (or a tunnel). Thus, various reasoning processes can be conducted when we detect overlapping object instances. SIGMA performs the detection of conflicting interpretations based on overlapping object instances.

Since SIGMA recognizes all possible objects by analyzing an image, some of the objects may occupy the same space. In order to decide which of such overlapping objects are allowed and which are not, we introduce *in-conflict-with* links. An in-conflict-with link connects a pair of object classes which cannot be located at the same place.

GRE examines all recognized object instances in the Iconic/Symbolic Database to detect mutually overlapping instances. Let I_1 and I_2 denote such instances and O_1 and O_2 their object classes, respectively. If there is an in-conflict-with link between O_1 and O_2, GRE establishes an in-conflict-with relation between I_1 and I_2, which implies that these instances are mutually conflicting.

Although GRE detects mutually conflicting instances, it does not make any decision on which of them is correct. Instead, it conducts independent reasoning based on each of the conflicting instances. That is, an instance can generate new hypotheses even if it is conflicting with another, and a new instance may be detected by analyzing a hypothesis. However, since hypotheses generated by mutually conflicting instances are also conflicting, they must not be accumulated. We will describe how such independent reasoning is realized in Section 3.4.3.

Our former image understanding system (Naga1980), when it found conflicting instances, immediately evaluated their reliability values and canceled the recognition of the less reliable one. In SIGMA, however, we consider that such an immediate decision can be wrong; since a conflict is detected at an intermediate stage of interpretation, the information based on which the decision is made is limited. That is, much new information (i.e., new object instances) may be obtained by continuing the interpretation process, as approach which allows a more reliable decision on which of the mutually conflicting instances is correct. In Section 3.1.4, we will discuss a method of selecting the most reliable interpretation from mutually conflicting interpretations.

In-conflict-with links have much to do with A-KIND-OF and PART-OF links. Let $O_{general}$ and $O_{specific}$ denote a pair of object classes connected by an A-KIND-OF link. In SIGMA, we consider that an instance of $O_{specific}$ is also an instance of $O_{general}$. This implies that when an instance of $O_{specific}$ is recognized, the corresponding instance of $O_{general}$ is automatically generated. Since these instances denote the same object, they are consistent even if they occupy the same space. In other words, when a pair of object classes is connected by an A-KIND-OF link, the establishment of an in-conflict-with link between them is prohibited.

On the other hand, when $O_{general}$ has two specialized object classes,

O_1 and O_2, instances of O_1 and O_2 should be regarded as conflicting if they are located at the same place; we consider a pair of specializations of $O_{general}$ to O_1 and O_2 as mutually exclusive. Thus, specialized object classes which are connected with the same general object class by A-KIND-OF links are always connected by in-conflict-with links. The same arguments hold for PART-OF relations.

In short, a pair of object classes which are connected by a sequence of $AKO_{specialization}$ and/or $PO_{whole-part}$ links alone are consistent and hence no in-conflict-with link can be established between them. On the other hand, a pair of object classes which are connected by a mixed sequence of $AKO_{generalization}$, $AKO_{specialization}$, $PO_{whole-part}$, and $PO_{part-whole}$ links are always connected by an in-conflict-with link. Thus, for simplicity, we will not illustrate such default in-conflict-with links in our world model.

Chapter 3

Algorithms for Evidence Accumulation

As we have described in Chaper 2, the SIGMA image understanding system consists of three cooperating reasoning modules plus the Question and Answer Module (QAM). The Geometric Reasoning Expert (GRE) performs evidence accumulation for spatial reasoning and constructs the interpretation of the scene. Often, GRE generates *hypotheses* about undiscovered objects and initiates the top- down verification analysis. A hypothesis is passed to the Model Selection Expert (MSE), which reasons about the most likely appearance of the object. Then, the description of the expected appearance is given to the Low-Level Vision Expert (LLVE), which verifies/refutes its existence in the image.

In this chapter, we describe the control structure of SIGMA, the representation of evidence in the Iconic/Symbolic Database, and detailed algorithms used in the evidence accumulation method by GRE. As for MSE, we mainly describe interactions between GRE and MSE, since we have not yet extensively studied algorithms for appearance model selection. The knowledge representation and reasoning method used in LLVE will be described in detail in Chapter 4.

In this chapter, we will describe practical algorithms to realize evidence accumulation for spatial reasoning, while Chapter 2 describes the process at the conceptual level. That is, at the conceptual level, object instances are active reasoning agents which perform reasoning autonomously using rules stored in corresponding object classes. At the realization level, on the other hand, they are static data structures representing various attributes of recognized objects. Based on stored attributes of instances, GRE evaluates rules to generate hypotheses and establish various symbolic relations between instances. In this sense, GRE is a rule evaluator which makes instances behave as specified in rules. The correspondence between these two levels of descriptions will become clear as we progress through this chapter.

We first present the overall control mechanism of SIGMA in Section 3.1. The detailed internal organization of GRE is given to describe its control structure. Section 3.2 describes the process of hypothesis generation and Section 3.3 the representation of evidence in the Iconic/Symbolic Database. (In what follows, we refer to the Iconic/Symbolic Database as simply *the database*.) In Section 3.4, we discuss detailed algorithms for examining the consistency among pieces of evidence in the database. Consistency examination is a kernel process performed by GRE. The consistency is examined from several different viewpoints. Section 3.5 describes the focus of attention mechanism, which selects the most important *situation* to be resolved. In Section 3.6, we discuss algorithms to resolve the selected situation. Various pieces of domain-specific knowledge are incorporated to resolve a situation. Sections 3.7 and 3.8 describe how such knowledge can be represented in terms of rules for spatial reasoning. Section 3.9 gives an illustrative example of the reasoning process by GRE.

3.1. OVERALL CONTROL STRUCTURE OF SIGMA

Figure 3.1 illustrates the overall analysis stages of SIGMA. At the very beginning of the analysis, SIGMA activates the initial segmentation process, which extracts image features corresponding to easily detectable object appearances. Based on the extracted image features, a group of *seed* object instances are generated and inserted into the database, from which the interpretation of the entire scene is constructed gradually.

At the second stage, SIGMA iterates the *interpretation cycle* to construct partial interpretations. During the construction, SIGMA establishes various symbolic relations among object instances in the database and activates the top-down analysis to detect missing object instances. When all construction activities terminate, SIGMA passes control to the QAM, by which various information about the constructed interpretation can be retrieved.

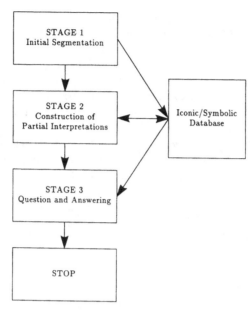

Figure 3.1.   Analysis stages in SIGMA.

3.1.1. Initial Segmentation Process

SIGMA starts its processing by initiating a search for easily detectable objects. This search is conducted by MSE.

Some objects in the scene have very prominent appearances, while others do not. Thus, it is reasonable to first detect objects with prominent appearances; image features corresponding to them can be easily and stably extracted even with simple image segmentation methods. On the other hand, the detection of objects with less prominent appearances requires accurate *a priori* information about their properties: brightness values, areas, locations, and so on. In high-resolution aerial images, for example, houses and roads are usually depicted as homogeneous regions of considerable area, and they have both characteristic shape properties and good contrast to their surroundings. On the other hand, driveways are very narrow and sometimes shadows are cast on them by trees. Therefore, we first detect houses and roads and then direct the search for driveways based on the information about already recognized objects.

Since the world model contains all types of object classed expected to appear in the scene, we first have to select those objects with prominent appearances. When SIGMA is initiated, MSE examines default attributes of object classes included in the world model and selects those object classes whose instances may be extracted stably. This selection involves the following two search processes in the world model:

1. *Selection of object classes with primitive shape information.* Since object classes at higher levels of PART-OF hierarchies have no primitive shape information, select those object classes at *leaf* nodes of the hierarchies. (Object classes without PART-OF links are also regarded as leaf nodes.)
2. *Determination of default appearances.* Since each selected object class may be connected to multiple specialized object classes (of different shapes) via A-KIND-OF links, first determine which specialized object class is most possible. Then, for each selected specialized object class, compute its default appearance in the image based on APPEARANCE-OF links. This process requires knowledge about various default properties of object classes and photographic conditions.

The output of this search process is described by a list of hypotheses describing constraints on the properties of image features expected to appear in the image. This list is called the *I-set* and is the goal of the initial segmentation by LLVE. MSE selects hypotheses from the I-set one by one and directs LLVE to extract image features which satisfy the associated constraints. For each extracted image feature, MSE instantiates the corresponding object class in the scene domain, and then inserts the object instance into the database. Such object instances are *seed* instances from which the interpretation of the entire scene is constructed.

In principle the above search process should be done whenever imaging conditions change or the world model is modified; easily extractable image features may change depending on photographic conditions (e.g., type of sensor, resolution, and size of the image). As will be described in Chapter 5, however, in our experiments we did not implement the search process but instead gave directly to the system a predetermined I-set. This is because (1) the world model used in the experiments was simple and fixed, and (2) all images to be analyzed were taken under similar photographic conditions. The set of hypotheses used as the I-set will be described in Section 5.3.

3.1.2. Interpretation Cycle of GRE

After the initial segmentation, control is passed to GRE, which starts constructing the interpretation based on the information stored in the database. Figure 3.2 illustrates the internal structure of GRE and its relations to MSE and LLVE. GRE iterates the following *interpretation*

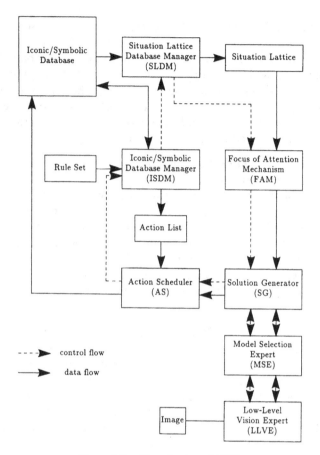

Figure 3.2. Organization of GRE.

cycle until no new information is added to the database. The interpretation cycle is illustrated by dotted lines in Fig. 3.2.

1. *Hypothesis generation.* At the first step of the interpretation cycle, each object instance stored in the database generates hypotheses about its related objects. This hypothesis generation process is realized as follows. First, for each object instance, the Iconic/Symbolic Database Manager (ISDM, Fig. 3.2) evaluates ⟨control-condition⟩ parts of the rules stored in the corresponding object class, and selects those rules which are applicable to that object instance. (Recall that rules represent the knowledge for spatial reasoning, and they are based on both *PART-OF* and spatial relations. See Section 2.4 for their structures and meanings.)

Then, ISDM evaluates ⟨hypothesis⟩ parts of the selected rules using properties of the object instance. By this evaluation, hypotheses are generated and inserted into the database. That is, ISDM is an evaluator of rules for spatial reasoning which evaluates ⟨control-condition⟩ and ⟨hypothesis⟩ parts of rules. [As will be described in (4), ⟨action⟩ parts are evaluated by another module in GRE.] Those rules used to generate hypotheses are recorded in the *action list* (Fig. 3.2).

2. *Evidence accumulation.* All pieces of evidence (i.e., instances and hypotheses) in the database are represented by using both iconic and symbolic data structures. We call an entity in the database representing an instance or a hypothesis a *database entity* (DE). At the second step of the interpretation cycle, the Situation Lattice Database Manager (SLDM) examines the consistency among DEs in the database and combines mutually consistent DEs into what we call a *situation*. The *situation lattice* in Fig. 3.2 is the data structure used to represent consistency relations among DEs.

3. *Focus of attention.* Next, the Focus of Attention Mechanism (FAM, Fig. 3.2) selects a situation to be resolved by examining the situation lattice. Since many consistent situations are usually formed by SLDM, FAM computes the reliability of each situation and selects the most reliable one. If there are several situations with equal reliability, FAM arbitrarily selects one of them. The selected situation is passed to the Solution Generator (SG, Fig. 3.2).

4. *Resolving the selected situation.* SG first computes a *composite hypothesis* based on the hypotheses included in the selected situation. Then, it examines if the situation contains a DE representing an object instance.

4-1. If the situation contains an object instance, SG broadcasts that instance to all *source* object instances which generated the hypotheses included in the situation. Each source object instance examines the returned instance and performs the *action* specified in the rule which was used to generate the hypothesis. Specifically, Action Scheduler (AS, Fig. 3.2) controls the execution of such actions. That is, AS evaluates ⟨action⟩ parts of those rules in the action list which were used to generate the hypotheses. Thus, like ISDM, AS is an evaluator of rules for spatial reasoning. Note that since many source instances are related to a situation, many ⟨action⟩s are to be evaluated to resolve the situation.

4-2. Otherwise, SG activates MSE to find an object instance which satisfies the constraints associated with the composite hypothesis. When a new object instance is returned from MSE, the same process as in (4-1) is performed. If MSE cannot find the required object instance, MSE returns *failure* to SG. Note that even in the case of failure, SG reports that result to all source instances; the source instances record the failure internally, and may activate rules to respond to such failure at the next interpretation cycle. At the realization level, AS evaluates ⟨action⟩ parts of rules in the action list using *failure* as a returned value, which may modify attributes of source instances. And new rules may become applicable due to such modifications.

GRE iterates the interpretation cycle until no new hypothesis is generated. At each interpretation cycle, one situation is selected and resolved, and the database is modified based on the solution of the selected situation. The modification of the database enables new rules to be activated, and new hypotheses and accordingly new situations are generated. Thus, the interpretation cycle is very similar to the *recognize-and-act cycle* used in production systems. Detailed algorithms used at each step in the interpretation cycle will be given in subsequent actions.

3.1.3. Specifying a Goal of the Analysis

The goal of the analysis by GRE is implicitly defined as follows:

Detect from the image all possible object instances which match the world model.

That is, GRE tries to create as many object instances as possible by examining all hypotheses generated. The interpretation cycle is terminated when no more hypotheses can be generated and every hypothesis has been verified/refuted.

The above goal (i.e., termination condition) is reasonable when we want to obtain the description of the entire scene. However, we sometimes want to detect only a specific object instance. In such cases, it is not efficient to wait until all object instances have been detected. Instead, we want to specify explicitly a goal of the analysis such as "find an object instance with specific properties." GRE should terminate the interpretation cycle immediately when such an instance is detected.

To support such explicit goal specification, a special built-in function, *STOP*, is included; whenever this function is evaluated, GRE terminates all activities of constructing the interpretation. A user can specify a goal by writing a rule whose action part contains this function.

Suppose, for example, the goal is to locate a road whose length is longer than 100 meters. This goal can be described by the following rule, which is stored in the *ROAD* object class:

⟨control-condition⟩: road.length > 100 meters

⟨hypothesis⟩: NIL

⟨action⟩: *STOP*

When a road instance longer than 100 meters is detected, the ⟨control-condition⟩ of this rule is satisfied. Since the ⟨hypothesis⟩ part of the rule is NIL, ISDM directly activates the ⟨action⟩ at the hypothesis generation step, which terminates the interpretation cycle. Then, GRE proceeds to the final question and answer stage (Fig. 3.1). That is, those rules whose ⟨hypothesis⟩ parts are NIL are used to perform special tasks. Sections 3.2 and 3.7 describe various types of tasks performed by such rules.

A goal specification can include requirements for relations between object instances. For example, the following ⟨control-condition⟩

and(road.length > 100 meters, number-of(road.along-house) > 2)

specifies that the interpretation cycle should be terminated when a road instance which is longer than 100 meters and is along at least two house instances is detected. Note that a goal is used only to specify the termination condition and that no goal-directed analysis is performed by GRE. (On the other hand, as will be described in the next chapter, LLVE performs the goal-directed image segmentation.)

3.1.4. Selection of the Final Interpretation

Interpretations constructed by GRE are described by *interpretation networks*. By an interpretation network, we mean a connected network consisting of mutually related object instances: recognized object instances are represented by nodes and relations among them (i.e., spatial relations and *PART-OF* relations) by arcs (Fig. 3.3). Note that although in SIGMA some object instances are linked by *A-KIND-OF* and *APPEARANCE-OF* relations, we can regard them as denoting the same object, and hence it is enough to discuss networks constructed by *PART-OF* and spatial relations alone. Some examples of interpretation networks including *A-KIND-OF* and *APPEARANCE-OF* relations will be given in Section 3.9.

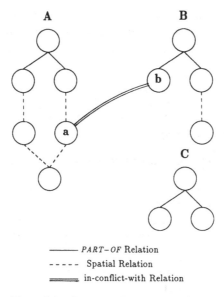

    —————— *PART-OF* Relation
    - - - - - Spatial Relation
    ═══════ in-conflict-with Relation

Figure 3.3. Interpretation networks (see text).

The following are general properties of interpretation networks:

1. All object instances in the same network are mutually *consistent*; spatial and *PART-OF* relations are established only between mutually consistent object instances. (The definition of consistency will be given in Section 3.4.) In Fig. 3.3, for example, we have three interpretation networks: A, B, and C.
2. As discussed in Section 2.4.4, multiple different object instances may be located at the same place in the scene. Usually, such instances are regarded as *conflicting* instances and an *in-conflict-with* relation is established between them. For example, nodes **a** and **b** in Fig. 3.3 represent conflicting instances, and networks A and B are regarded as conflicting interpretation networks. Since GRE establishes no new *PART-OF* or spatial relations between object instances belonging to mutually conflicting networks, no *PART-OF* and spatial relations are established across such networks. (We will discuss this property in detail in Section 3.4.3.)
3. Since some object instances may not have any relations to others, networks consisting of such instances are isolated from others, e.g., interpretation network C in Fig. 3.3. Note that neither consistency nor conflict is defined between mutually disjoint networks.

By an *interpretation,* we mean a set of nonconflicting interpretation networks. Since GRE constructs all possible interpretations, we need to select a good one as the final result. This selection can be done as follows:

1. Let *FINAL-INTERPRETATION* and *NETWORKS* be { } (i.e., empty set) and $\{N_1, \ldots, N_n\}$ (i.e., set of all constructed interpretation networks), respectively.

2. Iterate the following steps until *NETWORKS* becomes empty:

    2-1. Let N be the largest interpretation network among *NETWORKS*. Append N to *FINAL-INTERPRETATION*. The size of a network is measured by the number of constituent nodes.
    2-2. Eliminate all networks in-conflict-with N from *NETWORKS*.

In Fig. 3.3, for example, the interpretation networks A and C are regarded as the final interpretation.

The rationale behind this process is that a set of *correct* object instances usually satisfy spatial and *PART-OF* relations specified in the world model, so that they form a large interpretation network. In other words, we use the size of a network to evaluate the reliability of conflicting object instances. Let I_1 and I_2 denote mutually conflicting object instances (i.e., located at the same location), and N_1 and N_2 interpretation networks to which I_1 and I_2 belong, respectively. Suppose N_1 is larger than N_2. To include N_1 in the final interpretation means that we regard I_1 as a correct object instance and I_2 as an incorrect one. Thus, at the final stage of the analysis, the decision is made to determine which object instances are correct; during the interpretation process, no such decision is made and all detected object instances are regarded as possible interpretations.

Although most incorrect object instances are detected and eliminated by the above process, there may still remain some really incorrect object instances; if interpretation networks constructed from incorrect instances do not cause any conflict with the largest network, the incorrect instances are included in the final interpretation. For example, even if network C in Fig. 3.3 is incorrect in reality, we have no way to prevent it from being included in the final interpretation. Such incorrect instances could be removed by checking the amount of supporting evidence: that is, eliminate those instances without enough support. In order to implement such a process, we need to introduce a criterion based on which the reliability of each detected instance is evaluated.

In the implementation, we realize the above process of selecting the final interpretation by a question and answer process. That is, QAM was developed to answer various queries about constructed interpretation networks and to display the retrieved information iconically. After the interpretation cycle by GRE is terminated, QAM is activated. Given a query, QAM searches for specified object instances by examining properties of object instances and traversing arcs in interpretation networks. We can specify various conditions in a query: attributes of instances and various relational properties to other instances. All experimental results shown in Chapter 5 were obtained by using QAM.

3.2. HYPOTHESIS GENERATION

The first step in the interpretation cycle is hypothesis generation. At this step, ISDM (see Fig. 3.2) evaluates all rules that are applicable for each object instance so far detected.

Suppose I_0 is an instance of object class O. For each rule, say R_i, defined in O, ISDM evaluates its $\langle$control-condition$\rangle$. If the evaluation result is true, ISDM performs the following processings:

1. Evaluate the $\langle$hypothesis$\rangle$ part of R_i to generate a hypothesis and insert the generated hypothesis into the database.
2. Insert the $\langle$action$\rangle$ part of R_i together with its corresponding hypothesis into the *action list*, which records all the actions waiting for execution (Fig. 3.2).
3. Put a mark in I_0 representing that R_i has been applied, which prevents R_i from being applied again at later interpretation cycles. That is, once a rule is applied to an instance, it will never be applied again to the same instance.

Actions in the action list are called *delayed actions*. Each delayed action is associated with its corresponding hypothesis generated at step (1) above. Such hypothesis is called the *cause-of-delay* of the action.

For each instance in the database, ISDM evaluates the $\langle$control-condition$\rangle$ of every rule in the corresponding object class. However, this strategy is not efficient. A more efficient strategy would evaluate only those $\langle$control-condition$\rangle$s which are affected by modifications made to the attributes of the instance: once the $\langle$control-condition$\rangle$ of a rule is evaluated, it need not be evaluated again unless some attributes related to it are changed. Such an efficient evaluation strategy has been used in production systems to enhance performance (Newe1978, Forg1982).

Several points may be made regarding hypothesis generation. The first one concerns the limited generation of hypotheses. ISDM evaluates rules only for instances recorded in the database. This implies that only instances are allowed to generate hypotheses about related objects. Although this limitation could be relaxed, to do so would lead to the excessive generation of weak hypotheses—generated from hypotheses without the support of actual image features. (Section 2.3.1 discussed the logical implication of the limited generation of hypotheses.)

Another point concerns the bottom-up instantiation of object classes based on PART-OF relations. As discussed in Section 2.2.3, an instance of a part object class can directly instantiate its whole object class using the PART-OF relation between them. This bottom-up instantiation is realized by ISDM as follows.

When a new instance, I, of object class O is inserted into the database, ISDM examines its properties and determines if I is either *fully instantiated* or *partially instantiated*. Recall that object class O has a *kernel list* that specifies which part object classes of O must be instantiated in order for an instance of O to be fully instantiated. The kernel list of O consists of a set of sublists, each of which contains a group of part object classes of O connected by PART-OF links. ISDM compares the part object instances of I so far detected with each sublist in the kernel list, and if sufficient part instances of I have already been detected, it generates the parent instance of I based on the PART-OF relation. Thus, ISDM first examines a set of object instances newly generated at the last interpretation cycle, and performs bottom-up instantiation based on PART-OF relations. The rule evaluation for the hypothesis generation described above is done after this instantiation process.

ISDM performs several types of important tasks before generating hypotheses. The bottom-up instantiation described in the previous paragraph is one of them. IDSM performs these tasks guided by various types of rules stored in object classes.

As described in Section 3.1.3, the ⟨action⟩ of a rule whose ⟨hypothesis⟩ is NIL is not put into the action list, but instead is evaluated immediately by ISDM. Various tasks can be achieved by using such rules:

1. *Goal specification.* As described in Section 3.1.3, a goal of the analysis can be specified by using a rule whose ⟨action⟩ part contains a special function, STOP.
2. *Instance generation.* The bottom-up instantiation based on PART-OF relations described above is also implemented by using rules whose ⟨hypothesis⟩ parts are NIL. ⟨Action⟩ parts of instance generation rules contain a special built-in function MAKE-INSTANCE.

3. *Unification of duplicated instances*. As discussed in Section 2.2.3, duplicated descriptions of a whole object instance are often constructed during the reasoning process based on PART-OF relations. The unification of such duplicated instances is conducted by rules of this type. A special built-in function UNIFY-INSTANCES is designed for this unification. We will describe the process of the unification in detail in Section 3.6.4.

4. *Management of alternative hypotheses*. An instance often generates multiple hypotheses at first, and if one of them is verified, then it retracts the others. We call such hypotheses *alternative hypotheses*. The retraction of useless alternative hypotheses is realized by using rules of this type. A special built-in function REMOVE-ALL is designed to retract such hypotheses. The reasoning process using alternative hypotheses will be described in Section 3.8.

Illustrative examples of these rules will be given in Sections 3.7 and 3.8. In a sense, these tasks can be considered as preprocessings to propagate throughout the database side effects caused at the previous interpretation cycle. That is, before hypothesis generation at the $(n + 1)$th interpretation cycle, ISDM performs these tasks to complete the nth interpretation cycle.

In summary, at the conceptual level, object instances in SIGMA are regarded as active reasoning agents, and rules described in object classes specify their behaviors in various situations. To realize such object-oriented computation, ISDM and AS are implemented as evaluators of such rules, which make object instances behave as specified. In this sense, ISDM and AS can be regarded as interpreters of an object-oriented programming language. A user need not know about the detailed mechanisms of the interpreters, but has only to describe the knowledge about (behaviors of) objects in terms of rules. Various built-in functions can be considered as primitive vocabularies to describe the behaviors.

3.3. REPRESENTATION OF EVIDENCE

All hypotheses and instances are stored in the Iconic/Symbolic Database. In the database, both instances and hypotheses are represented by database entities (DE). Each DE is described by the following two parts:

1. *Iconic description*. This specifies a region which indicates the

location of a DE. The iconic description of a hypothesis specifies the locational constraint on the target object. As will be described in the next section, iconic descriptions play an important role in examining consistency among DEs. (Although we used 2D regions for aerial image understanding, 3D volumes should be used as iconic descriptions for image understanding of 3D scenes.)

2. *Symbolic description.* This describes attributes of and symbolic relations among DEs. Like an instance, a hypothesis is represented by a copy of the corresponding object class: a data structure with the same structure as that of the object class. Constraints on the target object associated with a hypothesis are represented by a set of linear inequalities with one variable (a slot name). Such constraints are filled into slots of the hypothesis.

Consider two DEs (hypotheses):

$$DE_1:$$
target-object = *HOUSE*
house.centroid = (100, 130)
230 < house.area < 300

and

$$DE_2:$$
target-object = *RECTANGULAR-HOUSE*
house.centroid = (100, 130)
250 < house.area < 320
house.region-contrast > 3

These descriptions of DE_1 and DE_2 are conceptual ones just for illustration. Their actual symbolic descriptions are as follows:

DE_1:
(and (= 'target-object '*HOUSE*)
    ((= 'house.centroid.x 100)(> = 'house.centroid.x 100)
    (< = 'house.centroid.y 130)(> = 'house.centroid.y 130)
    (< 'house.area 300)(> 'house.area 230))

and

DE_2:
(and (= 'target-object '*RECTANGULAR-HOUSE*)
      (< = 'house.centroid.x 100)(> = 'house.centroid.x 100)
      (< = 'house.centroid.y 130)(> = 'house.centroid.y 130)
      (< 'house.area 320)(> 'house.area 250)
      (> 'house.region-contrast 3))

Although some slots of DE_1 and DE_2 are not specified (e.g., shape properties), we can generate their iconic descriptions by using default values stored in the corresponding object classes. In the case of DE_2, for example, the default shape of *RECTANGULAR-HOUSE* is a rectangle whose length is twice as long as its width. In the case of DE_1, on the other hand, first the default appearance of *HOUSE* is selected as *RECTANGULAR-HOUSE* and then its actual shape is determined. Using these defaults, ⟨hypothesis⟩ parts of rules generate both iconic and symbolic descriptions for DEs representing hypotheses.

All generated iconic descriptions (i.e., regions) are mapped on the same 2D array representing the scene space, and the consistency between locational constraints is examined by checking overlaps between the regions.

It should be noted that while every hypothesis has both iconic and symbolic descriptions, some instances have no iconic descriptions. As discussed in Section 2.4, object classes are organized into hierarchical structures using *A-KIND-OF* and *PART-OF* links. In SIGMA, instances of those object classes at higher levels in the hierarchies are given no iconic descriptions: only those instances of the object classes at the leaf nodes of the hierarchies are allowed to have iconic descriptions. This is because if instances of higher-level object classes had iconic descriptions, they would always overlap with those of their child instances. This would introduce redundancy into the consistency examination.

Suppose, for example, *RECTANGULAR-HOUSE* is *A-KIND-OF HOUSE*. Then, every instance of *RECTANGULAR-HOUSE* is also regarded as an instance of *HOUSE*. Specifically, for each instance of *RECTANGULAR-HOUSE*, a corresponding instance of *HOUSE* is generated, and these instances are connected by an *A KIND-OF* relation in an interpretation network. Similarly, suppose *HOUSE* is *PART-OF HOUSE-GROUP*. Then, for each instance of *HOUSE*, an instance of *HOUSE-GROUP* representing a house group to which that instance belongs is generated (i.e., bottom-up instantiation based on a *PART-OF* relation, as discussed in Section 2.2.3). Since these pairs of instances (i.e., instances of *RECTANGULAR-HOUSE* and *HOUSE*, and those of

HOUSE and *HOUSE-GROUP*) are always consistent, we need not to examine their consistency during the interpretation process. As will be described in the next section, however, since the consistency examination is carried out for every pair of DEs whose iconic descriptions overlap, the consistency between such obviously consistent instances would be examined if all of them had iconic descriptions. Thus, only those instances of the most specialized object classes in *A-KIND-OF* hierarchies and the most primitive ones in *PART-OF* hierarchies are allowed to have iconic descriptions.

3.4. CONSISTENCY EXAMINATION AMONG EVIDENCE

The most important task performed by GRE is consistency examination among pieces of evidence in the database. In SIGMA, many duplicated pieces of evidence denoting the same object are created from the different object instances which are accumulated to realize a reliable analysis. The major role of consistency examination is to find such duplicated pieces of evidence. In other words, *consistency* between pieces of evidence implies that they denote the same object.

In GRE, the consistency examination is carried out by SLDM (Fig. 3.2). A pair of DEs are regarded as consistent if all of the following conditions are satisfied:

1. *Locational constraint.* Their iconic descriptions must intersect. It is also possible to impose some requirements on the size and shape of the intersecting area.
2. *Compatible attributes.* Here, attributes include both the object classes to which DEs belong and constraints associated with them. In order for DEs to be consistent, their object classes must be *compatible* and associated constraints must be *satisfiable*. The definitions of compatibility and satisfiability will be given in Section 3.4.2.
3. *Nonconflicting source instances.* An object instance which generated a hypothesis is called the *source instance* of the hypothesis. The source instance of an object instance is that instance itself. In order for DEs to be consistent, their source instances must not belong to conflicting interpretation networks.

In this section, we first describe an algorithm based on a lattice structure to compute the maximal set of consistent DEs. Then, we discuss consistency examination methods based on attributes and source instances, respectively.

3.4.1. Lattice Structure for Exminining Locational Constraint

A group of mutually consistent DE's are combined into what we call a *situation*. Such DEs are said to be *participating* in the formation of the situation. The *P-set* of a situation is defined as the set of DEs which participate in the formation of the situation. Situation S_a is *less than* situation S_b if the P-set of S_a is a subset of the P-set of S_b. This partial ordering is use to organize all the situations into a *situation lattice* (see Fig. 3.2). Note that a single DE is also a situation.

A pair of DEs are said to be *2-consistent* if they are consistent. In general, a set of n DEs is *n-consistent* if all possible $(n - 1)$ DEs in the set are $(n - 1)$-consistent. Clearly, a set of n DEs is n-consistent if and only if all possible pairs of DEs in the set are 2-consistent. Note that an n-consistent set of DEs is $(n - 1)$-consistent. For example, DE_1, DE_2, and DE_3 are 3-consistent if all of the following pairs are 2-consistent:

$$\{DE_1, DE_2\}, \{DE_2, DE_3\}, \{DE_1, DE_3\}$$

Figure 3.4 illustrates an example of the situation lattice. All DEs in the database are located at the bottom of the lattice, all situations consisting of pairs of 2-consistent DEs at the second level, and all situations consisting of n-consistent DEs at the nth level. Arcs in the lattice represent *less than* relations defined between situations. *Maximum consistent situations* are those situations which are not less than any other situations (Fig. 3.4). In what follows, we present algorithms to manage the situation lattice.

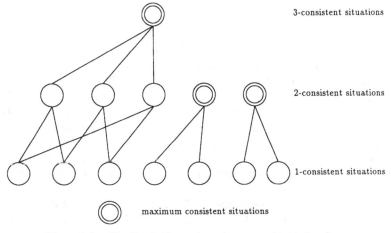

Figure 3.4. Situation lattice and maximum consistent situations.

Algorithm 3-1. Updating the Situation Lattice

Step 1: Insert DE_{new} at the bottom level of the situation lattice.

Step 2: Compute the set U of all DEs whose iconic descriptions intersect with that of DE_{new}. U contains all DEs which satisfy the locational constraint associated with DE_{new}. Note that DE_{new} itself is included in U.

Step 3: Examine the consistency between all pairs of DEs in U and compute all 2-consistent situations. Here, an n-consistent situation means a set of n mutually consistent DEs. Then, remove those situations whose P-sets do not contain DE_{new} from the computed situations. Let R denote the set of remaining situations. That is, R is a set of all 2-consistent situations including DE_{new}. The consistency here is examined based on both associated constraints and source instances of DEs. Detailed algorithms for this consistency examination will be described in Sections 3.4.2 and 3.4.3. Initialize $N = 2$.

Step 4: If R is empty, then exit. Otherwise, insert all elements of R at the Nth level of the situation lattice: new nodes representing situations in R are created and arcs are established based on *less than* relations among new inserted and old existing situations.

Step 5: Increment N by 1. Construct a set S of all N-consistent situations from the $(N - 1)$-consistent situations in R:

Set S = { } (i.e., empty set). For every pair of situations in R,
5-1. Compute the union of their P-sets. Let Q denote the union.
5-2. If $Q \in S$ or the size of Q is not equal to N, then do nothing.
5-3. If Q is not N-consistent, do nothing. It should be noted that an N-consistent situation is constructed from $N(N - 1)$-consistent situations. Therefore the N-consistency here can be checked simply by examining $(N - 1)$-consistent situations in the situation lattice.
5-4. Otherwise, append Q to S.

Set R = S.

Step 6: Go to Step 4.

When a new DE, say DE_{new}, is inserted into the database, SLDM updates the situation lattice by using Algorithm 3.1.

Figure 3.5 illustrates an example of how the situation lattice is updated when a new DE is inserted. In the figure, each DE is represented by a letter. A situation is represented by a string of letters representing all DEs in its P-set. Figure 3.5a shows the situation lattice and iconic descriptions of the DEs before inserting DE_E. Note that a pair of DEs (e.g., DE_C and DE_D) are not consistent even if their iconic descriptions are overlapping; sets of constraints associated with them are not consistent.

When DE_E are inserted, the set U computed at Step 2 in Algorithm 3.1 includes DE_A, DE_B, DE_C, DE_D, and DE_E. Suppose that DE_E is consistent with DE_A, DE_B, and DE_D. Then, the set R computed at Step

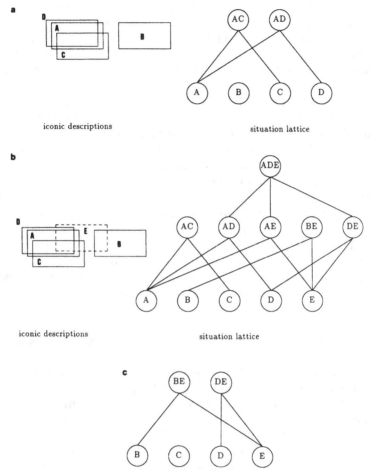

Figure 3.5. Modification of the situation lattice. (a) Situation lattice before inserting E. (b) Situation lattice after inserting E. (c) Situation lattice after removing A.

3 contains the following situations: $\{DE_A, DE_E\}$, $\{DE_B, DE_E\}$, $\{DE_D, DE_E\}$. These 2-consistent situations are inserted into the situation lattice at Step 4. After Step 5 is executed, R contains one 3-consistent situation $\{DE_A, DE_D, DE_E\}$; as illustrated in the situation lattice in Fig. 3.5a, DE_A and DE_B, and DE_B and DE_D are not consistent. The consistency of $\{DE_A, DE_D, DE_E\}$ can be easily verified by examining whether all of $\{DE_A, DE_D\}$, $\{DE_A, DE_E\}$, and $\{DE_D, DE_E\}$ are included in the situation lattice. Figure 3.5b shows the situation lattice after the updating process.

When a DE, say DE_{remove}, is being removed from the database, the situation lattice must be updated. This can be done simply by removing all the situations in the situation lattice which are larger than (i.e., including) DE_{remove}. Suppose, for example, that DE_A is removed from the situation lattice in Fig. 3.5b. Figure 3.5c shows the resulting situation lattice.

It is possible that the number of situations in the situation lattice may grow exponentially. In practice, however, this does not happen, because the number of DEs participating in a situation is usually quite small, e.g., two or three. That is, the height of the situation lattice is very low. This is because the number of rules which generate hypotheses of a specific object class is small and hypothesis generation is allowed only for object instances. In other words, the number of consistent DEs denoting the same object is limited. Moreover, although many new instances are inserted during the analysis, hypotheses are immediately removed once they are verified/refuted. Roughly speaking, when the number of DEs in the database is N, the space and time complexities of Algorithm 3.1 are both $O(N^2)$ spent at Step 3 to compute all 2-consistent situations. And N does not increase very much during the interpretation process.

3.4.2. Consistency Examination Based on Stored Attributes

At Step 2 in Algorithm 3.1, the locational consistency among DEs is examined by checking if their iconic descriptions overlap. The second step of the consistency examination is to check stored attributes of DEs. Attributes to be examined include object classes to which individual DEs belong and constraints associated with DEs.

Let DE_1 and DE_2 denote a pair of DEs satisfying the locational constraint, and let O_1 and O_2 denote the object classes of DE_1 and DE_2, respectively. In examining the consistency between DE_1 and DE_2, SLDM first examines the *compatibility* between their object classes (i.e., O_1 and O_2). That is, in order for DE_1 and DE_2 to be consisten, O_1 and O_2 must be *compatible*. In other words, if O_1 and O_2 are incompatible, then SLDM regards DE_1 and DE_2 as not consistent without examining their associated constraints.

Object classes O_1 and O_2 are said to be *compatible* if O_1 and O_2 are linked by an A-KIND-OF link, a PART-OF link, or a sequence of A-KIND-OF or PART-OF links. Each object class is compatible with itself. Specifically, when O_1 is A-KIND-OF (or PART-OF) O_2, O_2 is said to be *upper* than O_1. The relation *upper* is transitive. Then, the compatibility between O_1 and O_2

can be defined precisely as follows:

O_1 and O_2 are compatible iff one of them is upper than the other or
$O_1 = O_2$.

Consider the world model in Fig. 3.6. Suppose O_1 is RECTANGULAR-HOUSE and O_2 is HOUSE. Then, DE_1 and DE_2 can be consistent since O_1 is linked to O_2 by an A-KIND-OF link. Similarly, when O_1 is HOUSE-GROUP and O_2 is RECTANGULAR-HOUSE, DE_1 and DE_2 can also be consistent since O_1 is *upper* than O_2 (i.e., O_2 is linked to O_1 by a sequence of an A-KIND-OF link and a PART-OF link). On the other hand, when O_1 is RECTANGULAR-HOUSE and O_2 is L-SHAPED-HOUSE, O_1 and O_2 are incompatible and accordingly DE_1 and DE_2 are not consistent; neither of O_1 or O_2 is upper than the other.

If O_1 and O_2 are compatible, SLDM examines the *satisfiability* between the constraints associated with DE_1 and DE_2. Associated with every DE (i.e., instance or hypothesis) is a set of linear inequalities that constrain permissible values of the denoted object attributes. A simple constraint manipulation system is used to check the solvability of a set of inequalities. Given a pair of inequality sets (i.e., constraints), first intersect (i.e., take the union of) the inequality sets to generate the solution space, which is also represented by a set of inequalities. Then, examine the solvability of the solution space. If the solution space is

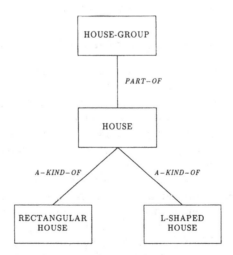

Figure 3.6. A world model for *HOUSE-GROUP*.

nonempty, then the constraints are regarded as *satisfiable*. This method is a simplified version of that used in ACRONYM (Broo1981).

If the constraints associated with DE_1 and DE_2 are satisfiable, SLDM proceeds to the final stage of the consistency examination, which checks the consistency based on the source instances of DE_1 and DE_2. Othewise, the DEs are regarded as inconsistent.

An important issue remains to be discussed regarding consistency examination based on attributes. Suppose that DE_1 and DE_2 are instances of O_1 and O_2, respectively, and that O_1 is at a level higher than O_2 (i.e., O_2 is a subpart or a specialization of O_1). Since the object classes of DE_1 and DE_2 are compatible, SLDM tries to examine the satisfiability between the constraints associated with them. However, since O_1 and O_2 are not the same object class, the constraints associated with DE_1 and DE_2 are described in terms of different terminologies; slot names of O_1 and O_2 are different. In order to examine the satisfiability between such constraints, we must translate the constraints into those described by the same terminology. This translation is done by using functions associated with the *PART-OF* and *A-KIND-OF* links connecting O_1 and O_2 (see Section 2.4.3). The translation is performed upward through the links. That is, in examining the satisfiability, first the constraints about the lower object class (i.e., O_2) are translated into those about the upper one (i.e., O_1), and then the satisfiability is examined based on the translated constraints. In the above example, the constraints associated with DE_2 are translated into those about O_1, and the satisfiability between the translated constraints and the constraints associated with DE_1 is examined.

3.4.3. Consistency Examination Based on Source of Evidence

Since GRE constructs all possible interpretations, a given interpretation may not be consistent with another. For example, a region in the image should not be interpreted as both house and road at the same time. We regard such interpretations as *conflicting* interpretations. Besides the consistency examination described above, SLDM detects mutually conflicting interpretations and performs an additional consistency examination based on supporting evidence.

As discussed in Section 2.4.4, a pair of object classes whose instances cannot be located at the same position are connected by an *in-conflict-with* link. SLDM examines the iconic descriptions of all object instances in the database and detects all pairs of overlapping object instances. Let I_1 and I_2 denote such a pair of overlapping instances and let O_1 and O_2 be their object classes, respectively. If O_1 and O_2 are connected by an in-conflict-with link, SLDM establishes an in-conflict-with relation between I_1 and I_2. Otherwise, no processing is performed. Note that not all pairs of object classes are connected by an in-conflict-with link; since shadows can be cast on any type of object, object class SHADOW is not connected with any other object classes via in-conflict-with links. Moreover, no in-conflict-with links are established between object classes connected via PART-OF or A-KIND-OF links. Recall that since a whole object instance (e.g., house-group) and a part object instance (e.g., house), which are connected by a PART-OF relation, occupy the same space, the former is given no iconic description to avoid redundant overlapping DEs in the database. This also holds for a general and a specialized object instance connected by an A-KIND-OF relation.

Let N_1 and N_2 denote the interpretation networks to which overlapping instances I_1 and I_2 belong, respectively. When there is an in-conflict-with relation between I_1 and I_2, we call N_1 and N_2 mutually conflicting interpretation networks. Since these two networks contain mutally conflicting information (i.e., I_1 and I_2), they must not be merged into one network. If a connected interpretation network were constructed from mutually conflicting (sub)networks (i.e., N_1 and N_2), it would involve an *internal conflict* and would be meaningless. Here by an internal conflict we mean that a pair of instances included in the same network are conflicting. In order to avoid the construction of such meaningless networks, spatial and PART-OF relations must not be established across mutually conflicting interpretation networks. That is, spatial and PART-OF relations can be established only between those object instances (i.e., nodes) belonging to nonconflicting interpretation networks.

Since spatial and PART-OF relations are established as the result of evidence accumulation, we need to assure that pieces of evidence generated based on mutually conflicting interpretations are not accumulated. For this purpose, the consistency examination described so far (i.e., one based on the locational constraint and constraints on object attributes) is not sufficient. We need to examine the consistency based on in-conflict-with relations established between interpretation networks.

Recall that the *source instance* of a piece of evidence refers to the instance which generates that evidence. The source instance of an instance is itself. Let DE_1 and DE_2 denote a pair of evidence and S_1 and S_2 their source instances, respectively. After examining the consistency between DE_1 and DE_2 based on the locational constraint and constraints on attributes, SLDM checks if S_1 and S_2 belong to mutually conflicting interpretation networks. If so, DE_1 and DE_2 are regarded as inconsistent. Otherwise, these two DEs are regarded as consistent and are accumulated to form a situation.

Consider the example shown in Fig. 3.7. Let S_1 be a house instance

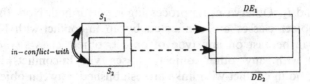

Figure 3.7. Hypotheses generated by conflicting instances.

and S_2 be a road instance. As discussed above, an in-conflict-with relation is established between these overlapping instances. Although S_1 and S_2 are mutually conflicting, they may generate hypotheses such that their iconic descriptions are overlapping and associated constraints are satisfiable. Let DE_1 and DE_2 be such hypotheses generated from S_1 and S_2, respectively. For example, DE_1 is a hypothesis about a neighboring house of S_1 and DE_2 that about a house along S_2. In this case, DE_1 and DE_2 cannot be consistent since their source instances, S_1 and S_2, belong to mutually conflicting interpretation networks.

Strictly speaking, the above consistency examination using source instances is not complete. For example, consider the case illustrated in Fig. 3.8a. DE_1 and DE_2 denote a house instance and a road instance belonging to the same interpretation network, respectively. Let DE_3 be a hypothesis about a driveway generated by DE_2, which overlaps with DE_1. Since the object classes of DE_1 and DE_3 are different, they are not considered as consistent and no further processing is performed. That is, DE_3 alone forms a situation. Suppose this situation is analyzed and an image feature satisfying the constraints associated with DE_3 happens to be extracted. Then, a database entity DE_4 representing the newly detected driveway instance is created (Fig. 3.8b). Since two instances of different object classes, DE_1 and DE_4, are overlapping, an in-conflict-with relation is established between them. Since these two object instances belong to the same interpretation network, an internal conflict is introduced into that network.

The current consistency examination algorithm cannot prevent networks with internal conflicts from being constructed as shown in the above example, and it is very difficult for the system to avoid them; the major reasons why internal conflicts are generated are inconsistency in the knowledge used for spatial reasoning and errors in segmentation. In the case shown in Fig. 3.8, for example, we can detect a potential internal conflict when we find that the hypothesis DE_3 generated from DE_2 is not consistent with DE_1 whereas DE_1 and DE_2 belong to the same network (i.e., these instances are consistent). If we established an in-conflict-with relation between DE_1 and DE_2 in Fig. 3.8a and prohibited DE_3 from

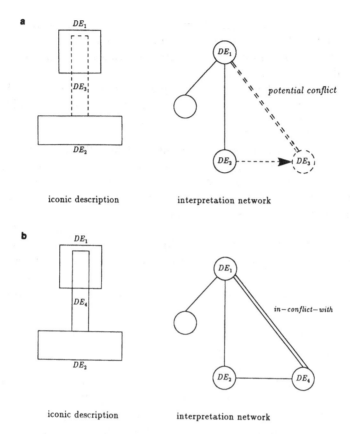

Figure 3.8. Introduction of an internal conflict. (a) Situation which may introduce an internal conflict. (b) Interpretation network with an internal conflict.

being analyzed, the internal conflict shown in Fig. 3.8b would be able to be avoided. However, we cannot determine the definite cause(s) of such potential internal conflict, based on which further reasoning is conducted:

1. Since the road instance DE_2 generates a hypothesis for driveway based on partial information so far obtained, much ambiguity is involved in the hypothesis DE_3, which makes it overlap (conflict) with DE_1. In this case, the description (i.e., constraints) of DE_3 should be modified.

2. Some already established relation or recognized instance in the interpretation network may be incorrect. For example, the relation between DE_1 and DE_2 (e.g., relation *FACING*) should not have been established, or one of them is an incorrect object

instance, or an erroneous image feature happened to be recognized. In this case, such relation or instance should be canceled.

3. DE_2 just generated DE_3 tentatively based on disjunctive knowledge (see Section 3.8): DE_2 tries to select one of the alternatives specified by the knowledge by generating hypotheses one by one corresponding to the alternatives. In this case, DE_2 should try the other alternatives.

Since the establishment of relations and hypothesis generation are performed based on the knowledge stored in individual object classes, it is very hard for the system to verify the overall consistency among such distributed knowledge sources. Moreover, the system has no way of reasoning about errors in segmentation.

In SIGMA, we assume that all knowledge stored in object classes is consistent and that no internal conflict is newly introduced once an interpretation network has been constructed. Under these assumptions, the consistency examination process described in this section is sufficient to assure that every interpretation network consists only of mutually consistent object instances.

3.5. FOCUS OF ATTENTION

The Focus of Attention Mechanism (FAM, see Fig. 3.2) selects the most reliable situation from the situation lattice. If there are several situations with equal reliability, FAM selects one arbitrarily. The situation selected by FAM is given to the Solution Generator (SG), which analyzes it to find the solution.

In this section, we describe the strategy for evaluating the reliability of a situation and discuss how the interpretation process can be controlled by using different reliability evaluation strategies.

3.5.1. Evaluation of the Reliability of a Situation

Given the situation lattice computed by SLDM, FAM selects the most reliable situation from the lattice:

1. Select a set of maximum consistent situations from the lattice.
2. For each selected situation, evaluate its reliability.
3. Pass the most reliable situation to SG.

In the current implementation, the evaluation of the reliability in (2) is done simply by summing up reliability values associated with DEs

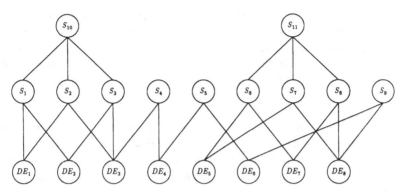

Figure 3.9. A situation lattice.

participating in each situation. No probabilistic model is involved in this evaluation process. In the lattice shown in Fig. 3.9, for example, there are five maximum consistent situations: S_4, S_5, S_9, S_{10}, and S_{11}. The reliability value of S_{10}, $R(S_{10})$, is defined as

$$R(S_{10}) = R(DE_1) + R(DE_2) + R(DE_3)$$

We will discuss how to define reliability values of hypotheses and instances [i.e., $R(DE_i)$] in Section 3.5.2.

As discussed in Section 1.3, the computation of reliability values is an important research topic in artificial intelligence (Gene1987). Especially in the area of expert systems, many computation schemes have been used, including the subjective Bayesian model (Duda76), the certainty factor (Shor1976), and the Dempster and Shafer theory (Shaf1975, Barn1981, Wesl1982).

The reasons why we did not use such schemes in SIGMA are as follows:

1. As discussed in Section 1.3, since the problem of image understanding is essentially underconstrained, we considered that it is reasonable to (a) first construct all possible interpretations in parallel based on the consistent (positive) information alone, and then (b) select the best one based on the overall reliability of each interpretation. Thus, multiple meaningful interpretations can be constructed for ambiguous figures like the Necker cube. And we can examine how each part of the image is interpreted in different interpretations.

In ordinary reliability computation schemes, when two pieces of evidence are mutually conflicting, the reliability of each piece of evidence is decreased to reflect the conflict. In our approach, on the other hand, such conflicting pieces of evidence are considered as alternative interpretations, based on which independent reasoning processes are conducted. In short, since SIGMA manipulates conflicting evidence independently, no negative information is incorporated into the reliability computation.

2. In ordinary reliability computation schemes, the set of entities for which reliability values are computed is fixed *a priori*: the total set of events in the Bayesian model, the frame of discernment in the Dempster and Shafer theory, and the set of "objects" to be labeled in relaxation labeling. In other words, ordinary reliability computation schemes assume that all entities involved in the computation are known *a priori* and hence that the purpose of the computation is just to calculate their reliability values.

In image understanding, however, new entities (i.e., image features and object instances) may be inserted at any time during the interpretation process. That is, new *a priori* unknown entities are dynamically introduced, a process which affects all reliability values so far computed.

Let us take relaxation labeling as an example. Let $\{IF_1, IF_2, \ldots, IF_n\}$ be a set of image features extracted at the initial segmentation. At first, reliability values of these image features are computed based on their attributes and mutual relations. When a new image feature is detected during the interpretation process, it is inserted into the set. Then, since the set of image features based on which reliability values were computed has been modified, the original reliability values become meaningless and hence the same process of reliability computation should be performed again based on the modified set. This recomputation should be done every time a new image feature is detected and each computation is time consuming. We encounter the same problem when the Dempster and Shafer theory is used for reliability computation; since probabilities are defined based on a fixed set of events (i.e., the frame of discernment), all computations should be done again from the first if the set is modified.

Since we have no definite idea about how to cope with this problem, we used the very simple reliability computation method described before. Note that a reliability computation based on intrinsic properties of each image feature can be performed without being affected by newly introduced image features. Thus, if we neglected the relational information to compute the reliability, we could use ordinary reliability computation schemes. In SIGMA, however, since a major source of the knowledge used for reasoning is spatial relations among objects, we cannot neglect the relational information.

3.5.2. Controlling the Interpretation Process

By assigning properly chosen reliability values to DEs, FAM can be used to control the direction of the interpretation process.

Recall that there are two different types of DEs in our system: instances and hypotheses. If a larger reliability value is assigned to an instance than to a hypothesis, a situation including an instance tends to be selected rather than one consisting only of hypotheses. Therefore, using such a reliability assignment rule, SIGMA first constructs partial interpretation networks by establishing relations among instances before trying to perform top-down object detection.

In fact, we assign a larger reliability value to an instance. This approach is followed based on the following observations:

1. Hypotheses are generated by applying rules to instances in the database, while an instance is directly supported by an observed image feature. It is resonable to assume that the information supported by the observation (i.e., instance) is more reliable than the information derived by the reasoning (i.e., hypothesis).
2. When instances are generated, they are isolated from the others, so that the information used for them to generate hypotheses is limited. Therefore, we should first establish relations among them to construct global interpretations so that new and more accurate hypotheses can be generated based on the established relations. The more accurate hypotheses become, the more precise the top-down image segmentation that can be conducted.

In the current implementation, we assign a unit reliability value to each hypothesis and five units to each instance uniformly. In principle, however, the reliability of each instance should be changed dynamically depending on the size of the interpretation network to which it belongs. That is, the fact that an instance belongs to a large network implies there are many other pieces of evidence consistent with that instance, and consequently that its reliability should be high.

It is natural to consider that the reliability of a hypothesis is proportional to that of its source instance. Therefore, situations including those hypotheses which are generated by instances belonging to large interpretation networks (i.e., those with high reliability values) are selected first. Thus, large networks grow fast by establishing relations to consistent instances and detecting new related instances by the top-down analysis, while small ones remain unanalyzed. That is, by assigning to instances reliability values proportional to the sizes of the interpretation networks to which they belong, we can realize a *feed-forward* analysis.

This control strategy is preferable to realize efficient interpretation. That is, we can incorporate the global FAM to select meaningful interpretations: once considerably large networks are constructed, we eliminate all small ones so as to reduce the number of DEs in the database and to avoid redundant computation:

1. SIGMA constructs all possible interpretations, so that the number of DEs tends to be very large and their manipulation requires significant computation time. Thus to eliminate redundant DEs saves much computation time.
2. Since small networks are usually constructed from incorrect object instances, the computation time spent in analyzing them is wasteful.

Note that this control mechanism may also eliminate correct interpretation networks; as discussed in Section 3.1.4, some correct networks may be isolated because the system cannot detect object instances which, if found, would bridge the networks to others. In the implementation, we did not use the global FAM described above, but we construct all possible interpretation networks and then eliminate inconsistent ones at the final stage.

3.6. RESOLVING A SELECTED SITUATION

Once a situation to be processed is selected, it is passed to SG. SG first computes a *composite hypothesis* from the hypotheses involved in the situation, and then activates MSE to verify the composite hypothesis. The result returned from MSE is then broadcast to all source instances related to the situation.

Section 3.6.1 discusses the computation of a composite hypothesis. Section 3.6.2 describes the reasoning process by MSE in searching for a solution. The broadcasting of the solution to the source instances is discussed in Section 3.6.3.

3.6.1. Constructing a Composite Hypothesis

A situation is a collection of consistent DEs. Given a situation, SG first computes a *composite hypothesis* which summarizes all constraints associated with the participating hypotheses. In what follows, we present some strategies for computing the composite hypothesis.

As discussed in Section 3.4, SLDM examines only the satisfiability among constraint sets associated with DEs participating in a situation. Thus, we need to define a strategy to actually compute a composite hypothesis based on participating hypothesis. Recall that hypotheses can be regarded as consistent even if their object classes are different.

One simple strategy is to regard the intersection of all constraints (i.e., inequalities) associated with hypotheses as the constraint set of the composite hypothesis. If object classes of participating hypotheses are different, that of the composite hypothesis is set to the most specialized and primitive one among those of the hypotheses. First, all constraints associated with the hypotheses are translated into constraints about such an object class by using functions attached to *A-KIND-OF* and *PART-OF* links, and then the intersection of the translated constraints is computed.

Suppose that we want to compute the composite hypothesis from DE_1 and DE_2 described below:

> DE_1:
> target object = *HOUSE*
> house.centroid = $(100, 130)$
> $230 <$ house.area < 300
>
> DE_2:
> target object = *RECTANGULAR-HOUSE*
> house.centroid = $(100, 130)$
> $250 <$ house.area < 320
> house.region-contrast > 3

Using the above strategy, the following composite hypothesis is computed:

> target object = *RECTANGULAR-HOUSE*
> house.centroid = $(100, 130)$
> $250 <$ house.area < 300
> house.region-contrast > 3

Another strategy is to use special rules to generate the description of a composite hypothesis. Special computations are performed to compose the composite hypothesis depending on the object classes of participating hypotheses. Suppose, for example, that two hypotheses about *ROAD* have constraints on road.orientation that differ by 90°. In this case, we may want to hypothesize a road junction and to compute the composite hypothesis for *ROAD-JUNCTION*. The construction of such a new type of hypothesis requires specialized knowledge.

We used the first strategy in SIGMA for simplicity. In general, however, some specialized knowledge is required to compute a composite hypothesis; hypotheses usually involve various constraints on geometric shapes, which require special computations to be composed.

3.6.2. Computing a Solution for the Selected Situation

SG activates MSE and passes a composite hypothesis as its goal. MSE searches for an object instance which satisfies the given goal.

First, MSE examines the P-set of the selected situation. If any instance is included in the P-set, then MSE returns that instance as the solution of the situation without performing any examination; the consistency of the DEs participating in the situation has been already examined by SLDM. Otherwise, MSE initiates the top-down analysis to detect a new instance satisfying the goal specification.

In the top-down analysis, MSE first examines the object class of the composite hypothesis:

1. If the object class is *primitive* (i.e., corresponding to a leaf node in *A-KIND-OF* and *PART-OF* hierarchies), the composite hypothesis is called a *primitive hypothesis*.
2. Otherwise, the composite hypothesis is called a *nonprimitive hypothesis*.

In the case of a primitive hypothesis, MSE first examines the *APPEARANCE-OF* links connected to the object class of the composite hypothesis, and selects the most plausible appearance of the target object. Then, MSE passes the selected appearance model to LLVE as its goal. If an image feature is returned from LLVE, MSE generates a corresponding object instance and returns it as the solution of the selected situation. Otherwise, MSE tries another appearance model, if any, and activates LLVE again. When no image feature can be extracted at all, MSE returns *failure* as the solution.

In the case of a nonprimitive hypothesis, MSE uses the knowledge associated with *PART-OF* and *A-KIND-OF* links to decompose the nonprimitive hypothesis into a set of primitive hypotheses. Let CH_a and O denote a composite hypothesis and its object class, respectively. Given CH_a, MSE conducts searching from O downward in the *PART-OF* (or *A-KIND-OF*) hierarchy until a primitive object class is found. During the search, MSE transforms the description (i.e., constraints) of a higher-level object class into that of a lower-level one by using a translation function associated with a *PART-OF* (or *A-KIND-OF*) link (see Section 2.4.3).

When the description of a primitive object is obtained, it is passed to LLVE as its goal, and the same process as in the first case is executed. When no image feature is extracted by LLVE, MSE searches for the next possible primitive object class and repeats the same process until an instance of some primitive object class *lower than* O is detected.

Since there may be many ways in which a nonprimitive hypothesis can be translated into a primitive one, we store the knowledge to guide the search in each object class. With each nonprimitive object class in a *PART-OF* hierarchy we associate a *decomposition strategy*. It is represented as an ordered list which describes in what order the object classes at the lower level (i.e., part object classes) should be examined. The same knowledge, named the *specialization strategy*, is incorporated to guide the search in an *A-KIND-OF* hierarchy. For example, suppose that *HOUSE* is connected to *RECTANGULAR-HOUSE* and *L-SHAPED-HOUSE* by *A-KIND-OF* links. One may specify the specialization strategy of *HOUSE* by (*RECTANGULAR-HOUSE*, *L-SHAPED-HOUSE*), which means that a rectangular house is more likely to be present than an L-shaped house.

3.6.3. Solution Broadcasting

The solution detected by MSE is passed to the Action Scheduler (AS, see Fig. 3.2) via SG. Using the solution provided by SG, AS schedules and activates actions in the action list. Recall that actions in the action list correspond to ⟨action⟩ parts of the rules which were used to generate hypotheses (see Section 3.2). Two possible types of solutions may be provided to AS: (1) *failure*: the composite hypothesis cannot be verified, or (2) an instance which satisfies the composite hypothesis.

In both cases, AS selects those actions in the action list whose *causes-of-delay* are included in the P-set of the resolved situation. Let the solution be I_0, the list of the selected actions $(A_1, \ldots, A_n)$, and the causes-of-delay of A_i $(i = 1 - n)$ H_i, respectively. AS performs the selected actions one by one sequentially:

1. Select one action, A_i, from $(A_1, \ldots, A_n)$.
2. Replace all the references to H_i in A_i by I_0.
3. Evaluate A_i.
4. Remote H_i from the database.

(We will discuss what processing can be performed by actions in more detail in Section 3.7.)

Note that the above process is executed even if the solution is *failure*; as will be described later, the information that a hypothesis has been refuted (i.e., the solution is *failure*) is also very useful for its source instance to perform further reasoning.

In the implementation, no explicit scheduling (based on some priority) is done to order the selected actions: AS examines the action list from the beginning and selects all executable actions. Since actions can cause various unrecoverable side effects in the database, the environments in which an action is evaluated change depending on the order in which the selected actions are evaluated. In most cases, however, the same final result is obtained after all actions are evaluated, whereas different intermediate states are produced depending on the order of the evaluation. This is because side effects caused by actions are usually to establish relations between source instances of a selected situation and the consistency between such instances has already been verified.

The removal of hypotheses from the database has the following side effect. If a hypothesis, say H_0, is removed from the database, then all the situations in the situation lattice whose P-sets contain H_0 are also removed (see Fig. 3.5c).

Note that to remove an existing hypothesis is a commitment made by the system. Suppose instance I_1 generated hypothesis H_1 and another instance I_2 satisfies H_1. By solving the situation whose P-set is $\{H_1, I_2\}$, relation REL is established between I_1 and I_2 and H_1 is removed from the database. Since the hypothesis is removed, I_1 cannot examine if still another instance, say I_3, can satisfy REL: I_2 is actually incorrect and the correct one (i.e., I_3) is detected later. If H_1 were preserved even after I_2 is detected, I_1 could recognize that it should establish REL to I_3 rather than I_2.

However, to keep all generated hypotheses causes a very serious problem in managing the database; its size and computation time increase explosively. Moreover, as will be described in Section 3.8, some hypotheses are generated only tentatively to examine environments of an instance, so that they should be removed once information about the environments is obtained. These are the reasons why we remove hypotheses immediately when they are verified/refuted.

When all the selected actions are evaluated, AS terminates and an interpretation cycle is completed. Then, the next cycle is initiated from the hypothesis generation process.

3.6.4. Unifying Partial PART-OF Hierarchies

In SIGMA, an instance of each part object class generates its parent instance independently based on a PART-OF relation. Since a whole object class is linked to many part object classes, multiple instances of a whole object class denoting the same entity are generated from those of its part instances. Thus, the process of unifying partially instantiated *PART-OF* hierarchies is a quite common action executed in SIGMA. (See Figs. 2.7 and 2.8 and the corresponding text in Section 2.2.3 for an explanation of how this unification process is executed.) In what follows, we will describe the detailed algorithm for unifying a pair of *PART-OF* hierarchies. (We will use *unification* and *merging* interchangeably.)

There are two possible realizations of the unification process:

1. Use a rule as follows:

   ⟨control-condition⟩: {examine if an instance has multiple parents}
   ⟨hypothesis⟩: NIL
   ⟨action⟩: {examine the consistency of the parents and if consistent unify them}.

   This rule is stored in a part object class. It is evaluated by ISDM at the beginning of a new intepretation cycle, since the ⟨hypothesis⟩ part is NIL. That is, multiple *PART-OF* relations to a part instance were established in the last interpretation cycle.

2. Use a rule as follows:

   ⟨control-condition⟩: {condition for top-down hypothesis generation}
   ⟨hypothesis⟩: {procedure to generate a hypothesis of a part object class}
   ⟨action⟩: {establish a *PART-OF* relation to a returned part instance}
      IF {the detected part instance has multiple parents}
      THEN {examine their consistency and if consistent unify them}.

   This rule is stored in a whole object class. The unification is executed by AS, since its process is described in the ⟨action⟩ part. That is, the ⟨action⟩ part performs the unification after establishing a *PART-OF* relation in the same interpretation cycle.

In principle, there is no essential difference between the reasoning processes executed by these two methods. We can use one of them by writing rules in corresponding object classes. In this section, we will describe the unification process based on the first method, because the detailed process of unification can be highlighted. On the other hand, the second method is used in the system itself and in the illustrative example in Section 3.9.

Let I_1 be an instance of a whole object O_{whole} and O_{part} be a part object class of O_{whole}. Note that a rule of type (1) above is stored in O_{part}. Suppose a hypothesis was generated by applying a rule in O_{whole} to I_1, which performs top-down hypothesis generation based on the PART-OF relation between O_{whole} and O_{part}. This rule itself is the same one as those used for reasoning based on spatial relations, and the ⟨action⟩ part of the rule just performs the following processing:

1. If the solution given by SG is not *failure*, establish a PART-OF relation to the instance, say I_0, returned as the solution.
2. Then establish a (reverse) PART-OF relation from I_0 to I_1.

The action is finished at this stage in this interpretation cycle.

At the beginning of the next interpretation cycle, a rule stored in O_{part} is appied to I_0 to check if I_0 has multiple parent instances [i.e., a rule of type (1) above]. If so, ISDM immediately evaluates the ⟨action⟩ part of that rule, since its ⟨hypothesis⟩ part is set to NIL. The ⟨action⟩ part contains special functions to examine the consistency between multiple parent instances and unify them. (An illustrative example of such a rule will be given in the next section.) The unification process includes the generation of a new parent instance, the removal of old parent instances, and the establishment of PART-OF relations to the new parent instance (Figs. 2.7 and 2.8).

In order to determine if two parent instances are consistent, some special knowledge is required. We store a special function in each parent object class, by which the consistency of a pair of instances of that object class is examined. A rule for unification stored in a part object class uses this function.

The result of the consistency examination (i.e., a value returned from a special function) is one of the following:

1. *Succeed*: parent instances are consistent and denote the same thing.
2. *Fail*: parent instances are mutually conflicting.
3. *Inconclusive*: not enough information is available to determine the consistency.

In the first case, parent instances are merged into one instance. A special function to merge them is stored in the ⟨action⟩ part of the rule.

In the second case, a copy of the shared part instance is newly generated to separate conflicting interpretations (i.e., parent instances). For example, let I_1 and I_2 be parent instances which share a part instance

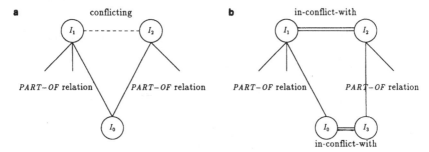

Figure 3.10. Separating an interpretation network. (a) Before separation. (b) After separation.

I_0 and are mutually conflicting (Fig. 3.10a). Then, a new instance I_3 is generated by copying I_0. After removing a PART-OF relation between I_2 and I_0, I_3 and I_2 are connected by a PART-OF relation. Then, in-conflict-with relations are established between I_1 and I_2, and I_0 and I_3 (Fig. 3.10b). Thus, after this processing a pair of conflicting interpretation networks are constructed. Note that this case rarely happens; the fact that parent instances sharing a part instance are mutually conflicting means the knowledge used by them for spatial reasoning is not consistent. In fact, we did not encounter any example of this case in our experiments.

The third case is common in SIGMA. It indicates that some information (slot values) which is essential to the consistency examination has not been obtained yet. In this case, the unification process is delayed until the required information becomes available.

Figure 3.11 illustrates an example of the third case. HOUSE-GROUP is defined as a group of regularly arranged HOUSEs which face the same side of the same ROAD. This means that it is crucial for a house group instance to establish a spatial relation to a road instance. In other words, even if a group of house instances are arranged regularly, it cannot be organized into a house group instance unless all member house instances face the same road instance.

Suppose that, as shown in Fig. 3.11b, a hypothesis generated by HG_1 is consistent with a member house H_3 of HG_2, and a new PART-OF relation is established between HG_1 and H_3: two house group instances (i.e., HG_1 and HG_2) share a house instance (i.e., H_3). Then, H_3 activates the consistency examination between HG_1 and HG_2. The result of this consistency examination depends on whether or not HG_1 and HG_2 have established spatial relations to their facing roads. If both house group instances have established spatial relations to the same road instance,

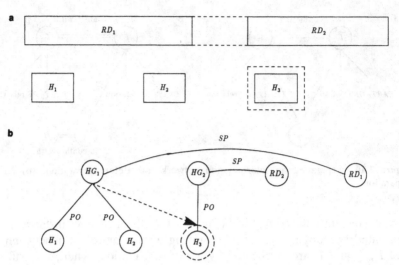

Figure 3.11. Situation which generates a frozen *PART-OF* relation. (a) Iconic description. (b) Interpretation network.

then they are consistent and are merged into one. On the other hand, if either of them has not established the relation, it is not possible to verify the consistency between them; HG_1 and HG_2 may face different roads. Moreover, even if two house group instances have established relations to different road instances, it is still possible for them to be consistent, because those road instances may be merged into one later. Figure 3.11 shows this situation. Thus, in the latter two cases, the result of the consistency examination becomes inconclusive.

If the result of the consistency examination is inconclusive, a new *PART-OF* relation from a parent instance to a shared part instance is frozen. In the case shown in Fig. 3.11, the *PART-OF* relation between HG_1 and H_3 is frozen. The system records all frozen relations and their causes in a common place. Such information is returned from the consistency examination process. A frozen relation can be reactivated if its cause is resolved by analyzing another situation. That is, after resolving a situation, causes of frozen relations are examined to see if any of them can be reactivated as a result of resolving that situation. In Fig. 3.11, for example, if a pair of road instances, RD_1 and RD_2, are merged into one instance, the frozen relation from HG_1 to H_3 is reactivated. Then, a pair of house group instances, HG_1 and HG_2, become consistent and consequently are merged into one. All these processes of freezing and reactivating relations are performed by ISDM at the beginning of a new interpretation cycle.

Even if the relation from HG_1 to H_3 is frozen, HG_1 can participate in other situations. That is, although HG_1 cannot use the frozen relation, it can generate hypotheses based on other relations and its attributes. In the current implementation, if there remain frozen relations even after all analysis is finished, they are discarded as nonverifiable evidence. Another possible strategy of managing such remaining frozen relations is to make a copy of each shared part instance and to generate disjoint interpretation networks as was done in Fig. 3.10. In such a strategy, however, it is difficult to decide whether or not an in-conflict-with relation is to be established between the generated interpretation networks, because no real conflict has been detected between them.

3.7. TAXONOMY OF ACTIONS

In this section, we discuss a taxonomy of actions that are often executed by rules for spatial reasoning. Actions here refer to procedures (functions) described in ⟨action⟩ parts of rules.

The most common type of action is to fill in slots (attributes) of an instance. For example, consider the following rule, which is stored in *HOUSE-GROUP*:

> *Rule for reasoning based on a spatial relation*:
> ⟨control-condition⟩: (conditions)
> ⟨hypothesis⟩: $H = AR(\text{self}, ROAD)$
> ⟨action⟩: self.along-road $= H$

This rule specifies the following knowledge:

1. Generate a hypothesis for *ROAD* if (conditions) are satisfied by a *HOUSE-GROUP* instance, say HG_1. *AR* denotes a function to generate the hypothesis, to which the source instance (i.e., HG_1) and the object class of the hypothesis are passed as arguments. Recall that *self* is a special variable, which is bound to an object instance to which the rule is applied. We can use various predicates as (conditions): those for examing attributes like area size and location and those for checking relations to other instances like the number of constituent house instances.
2. Then, if the target *ROAD* instance, say R_1, is detected, it is bound to variable H.
3. Finally, store the ID number of (i.e., pointer to) R_1 into the specified slot of HG_1. That is, the slot *along-road* of *HOUSE-GROUP* is used to store a symbolic spatial relation to a related *ROAD* instance.

As described before, actions sometimes create new instances directly (see Section 3.2) or unify multiple instances (see Section 3.6.4). Such actions are described by using the following two special built-in functions:

MAKE-INSTANCE: Create an instance and insert it into the database
UNIFY-INSTANCES: Unify a group of instances into a single instance

For example, the following rule is stored in *RECTANGLE* to generate an instance of *RECTANGULAR-HOUSE*:

> *Rule for instantiation*:
> ⟨control-condition⟩: *IS-RECT-HOUSE*(self)
> ⟨hypothesis⟩: NIL
> ⟨action⟩: *MAKE-INSTANCE*(*RECTANGULAR-HOUSE*, *TRANS*(self))

This rule describes the following piece of knowledge: If a *RECTANGLE* instance which satisfies the *IS-RECT-HOUSE* predicate is created, then create a *RECTANGULAR-HOUSE* instance and insert it into the database.

Function *IS-RECT-HOUSE* is a predicate to examine if a *RECTANGLE* instance can be regarded as *RECTANGULAR-HOUSE*. Function *TRANS* transforms descriptions of *RECTANGLE* into those of *RECTANGULAR-HOUSE*; since *RECTANGLE* and *RECTANGULAR-HOUSE* are object classes in the image and scene domains, respectively, they are described in terms of different terminologies. A major function of *TRANS* is to perform the geometric transformation between the image and scene coordinate systems. In short, this rule together with *TRANS* defines the semantics of the *APPEARANCE-OF* link between *RECTANGULAR-HOUSE* and *RECTANGLE* (see Section 2.4.3).

It should be noted that rules such as those above, which instantiate object classes in the scene domain based on subject instances in the image domain, are evaluated by MSE, That is, when instances in the image domain (e.g., rectangles) are generated by LLVE at the initial segmentation or in the top-down analysis, MSE uses such rules to generate corresponding instances in the scene domain (e.g., instances of *RECTANGULAR-HOUSE*).

Knowledge about the bottom-up instantiation based on *PART-OF* relations can be described in a similar way by using *MAKE-INSTANCE*. Let I_1 be an instance of O_{part}, which is a part of O_{whole} (i.e., O_{whole} and O_{part} are connected by a *PART-OF* link). The following rule, by which I_1 instantiates an instance of O_{whole}, is stored in O_{part}

> *Rule for bottom-up instantiation*:
> ⟨control-condition⟩: *FULLY-INSTANTIATED*(self, O_{part})
> ⟨hypothesis⟩: NIL
> ⟨action⟩: *MAKE-INSTANCE*(O_{whole}, *FUNCT*(self))

Predicate *FULLY-INSTANTIATED* examines if I_1 is fully instantiated by comparing its already computed attributes with the kernel list of O_{part}: for example, if O_{part} is a leaf node in a *PART-OF* hierarchy, its kernel list is (( )), so that *FULLY-INSTANTIATED* always evaluates true, and the rule is immediately applied to I_1. Function *FUNC* computes attributes of an instance of O_{whole} to be generated based on those of I_1. That is, it is nothing but the function $F_{part-whole}$ described in Section 2.4.3, which is associated with the *PART-OF*$_{part-whole}$ link from O_{part} and O_{whole}. In short, this rule defines the semantics of that *PART-OF*$_{part-whole}$ link.

As for the unification of instances, consider the following piece of knowledge:

If more than one *HOUSE-GROUP* instance is filled in the *belongs-to* slot of a *HOUSE* instance, examine the consistency between such multiple instances, and if they are consistent, unify it. (The slot *belongs-to* is used to store a symbolic *PART-OF* relation from a part object to its whole.)

This knowledge for unifying multiple parent instances can be described by the following rule, which is stored in object class *HOUSE*:

Rule for unifying multiple parent object instances:
⟨control-condition⟩: *NUMBER-OF-ELEMENTS*(self.belongs-to) > 1
⟨hypothesis⟩: NIL
⟨action⟩: *UNIFY-INSTANCES*(self.belongs-to, *COMBINE-H*(self.belongs-to))

Function *COMBINE-H* examines the consistency between multiple parent object instances, and if they are consistent, returns a merged parent instance: generate a new parent instance and remove the original ones. *UNIFY-INSTANCES* performs several of the types of processing described in Section 3.6.4, depending on the value returned by *COMBINE-H*: succeed, fail, or inconclusive. At the beginning of every interpretation cycle, this rule is applied to each instance of *HOUSE* to eliminate multiple *PART-OF* relations connected to a single *HOUSE* instance.

Instances in SIGMA usually generate hypotheses even if little information is available. This process of active hypothesis generation is useful to perform reasoning with incomplete information. However, it is possible that those hypotheses generated based on little information are not correct and will consequently be refuted. Actions which deal with such refuted hypotheses are also important. Basically there are two possibilities to manage refuted hypotheses: (1) modify the attributes of (constraints associated with) a refuted hypothesis, or (2) retract a refuted hypothesis and generate a different one. In what follows, we describe actions for (1). Those for (2) will be described in the next section; actions for (2) are performed based on disjunctive knowledge, a concept which should be discussed extensively.

Usually, a hypothesis is removed from the database by AS after SG proposes a solution for it. On the other hand, when a hypothesis is refuted or a solution is not acceptable, we may want to update attributes of the original hypothesis and retry its verification, since new information has since been obtained. That is, since an instance generates many hypotheses at the same time, solutions to some hypotheses are returned before a solution to a specific hypothesis is computed. Since those returned solutions increase the information contained in the instance, it can generate a more accurate hypothesis using such increased information.

Function *UPDATE* is intended to describe the process of updating attributes of a hypothesis. For example, consider the following rule:

> *Rule for updating attributes of a hypothesis*:
> ⟨control-condition⟩: (conditions)
> ⟨hypothesis⟩: $H = F(\text{self})$
> ⟨action⟩:
>   IF *NOT-ACDEPTABLE*(H) then *UPDATE*(H, CS_1)
>   ELSE...

The action specifies that if the solution proposed for H is not acceptable, then replace some attributes of H by CS_1 (newly obtained information). By evaluating ⟨action⟩ parts as above, AS conducts the updating process. Otherwise it removes verified/refuted hypotheses from the database.

3.8. REASONING BASED ON DISJUNCTIVE KNOWLEDGE

As discussed in Section 2.3.6, most of the knowledge used for spatial reasoning in SIGMA is positive and conjunctive. While no negative knowledge is used in SIGMA, we can realize spatial reasoning based on disjunctive knowledge by using *metarules*: rules which control the evaluation of other rules.

Consider the following piece of knowledge, which includes disjunction:

Knowledge 1:
   A *ROAD* which faces a *HOUSE-GROUP* is located at either the left or right side of the *HOUSE-GROUP* axis.

This knowledge can be described by the following logical expression:

$$\forall x[\textit{HOUSE-GROUP}(x) \supset \exists y[\{\textit{ROAD}(y) \land \{\textit{FACE\_RIGHT}(x, y)$$
$$\oplus \textit{FACE\_LEFT}(x, y)\}]] \quad (3.1)$$

Here $\oplus$ denotes exclusive OR, which implies that a *HOUSE-GROUP* can face only one *ROAD*. Note that we eliminated the uniqueness constraint in the above expression (see Section 2.3.1). Another example of disjunctive knowledge is as follows:

Knowledge 2:
    A *ROAD* is connected to either a neighboring *ROAD-PIECE* or an *INTERSECTION*, or terminates at a *DEAD-END*.

As is obvious from these examples, it is not rate that we have to conduct reasoning based on disjunctive knowledge. In SIGMA, we assume that all disjunctions imply exclusive OR: case A or case B means case A $\oplus$ case B.

In general, reasoning based on disjunctive knowledge involves a decision process to determine which case specified by the knowledge holds. In image understanding, however, the information required for such a decision process is not *a priori* available. Thus we first of all have to gather information and then decide which case holds in the scene under analysis. Based on this principle, we implemented the following two strategies in SIGMA to realize spatial reasoning based on disjunctive knowledge:

Strategy 1:
1. First, select the most plausible case and generate a hypothesis based on the knowledge specified in the selected case. The generated knowledge is called a *preferred hypothesis*. In the case of knowledge 2 above, a *ROAD* instance first generates a hypothesis for *ROAD-PIECE* since usually a *ROAD* can be extended further by merging with a neighboring *ROAD-PIECE*.
2. If the preferred hypothesis is successfully verified, then do nothing for the other cases.
3. Otherwise, retract the hypothesis and generate the next preferred one based on the knowledge. In the next example, a *ROAD* instance generates a hypothesis for *INTERSECTION*.

Strategy 2:
1. Generate multiple hypotheses at the same time corresponding to all possible cases specified by disjunctive knowledge. These

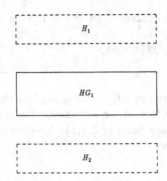

Figure 3.12. Alternative hypotheses for ROAD.

hypotheses are called *alternative hypotheses*. In the case of knowledge 1 above, a pair of alternative hypotheses about ROAD are generated from a HOUSE-GROUP instance (Fig. 3.12); the instance has no preference about the location of its facing road.

2. When one of the alternative hypotheses is verified, remove all the others.

The first strategy incorporates so-called *failure-driven analysis*. It is similar to the sequential depth-first search with backtracking used in Prolog. When knowledge about which case is most plausible is available, this strategy is effective. In failure-driven analysis, the preference of possible cases can be dynamically changed by analyzing the cause of a failure: the next possible case to try is determined based on the cause of a failure. Such dynamic selection of possible cases increases the flexibility and effectiveness of the analysis. In software engineering, such a control mechanism is called *intelligent backtracking* (Cox1981, Cox1986). In the above ROAD example, if a hypothesis for ROAD-PIECE is refuted and the cause of the failure is that its expected region contains green areas, then a hypothesis for DEAD-END should be generated rather than one for INTERSECTION.

The implementation of the first strategy does not require any new mechanism. As described in the last example in the previous section, when a solution returned for a hypothesis is not satisfactory, we can retract it and generate a new one by writing such processes in an ⟨action⟩ part of a rule. Function UPDATE is used for this processing, since it can modify any attribute of a hypothesis: object class, iconic description, and constraints. In short, disjunctive knowledge is encoded in procedures stored in rules.

When no knowledge about the preference among possible cases is available, the first strategy may be inefficient and the second one can be used. The second strategy involves parallel application of rules and requires a new mechanism to control (synchronize) the rule evaluation.

Suppose a *HOUSE-GROUP* instance HG_1 is going to generate a pair of alternative hypotheses about *ROAD*, H_1 and H_2 (Fig. 3.12). The generation of each alternative hypothesis should be done by a different rule since different functions are used in $\langle$hypothesis$\rangle$ and $\langle$action$\rangle$ parts to perform reasoning in different cases. Thus, we store the following two rules in *HOUSE-GROUP*, which generate H_1 and H_2, respectively:

> *Rule R_1:*
> $\langle$control-condition$\rangle$: (condition 1)
> $\langle$hypothesis$\rangle$: $H_1 = F_l(\text{self})$
> $\langle$action$\rangle$: self.along-road $= H_1$
>
> *Rule R_2:*
> $\langle$control-condition$\rangle$: (condition 2)
> $\langle$hypothesis$\rangle$: $H_2 = F_r(\text{self})$
> $\langle$action$\rangle$: self.along-road $= H_2$

These rules are the same as those used for ordinary spatial reasoning, and no disjunctive knowledge is represented by them. The parallel application of the rules can be realized by setting (condition 1) = (condition 2). Moreover, since rules are applied when little information is available, their $\langle$control-condition$\rangle$ parts are usually set as *true*: whenever an instance of *HOUSE-GROUP* is generated, these rules are applied.

We use the following *metarule* to control the evaluation of R_1 and R_2. This rule is also stored in *HOUSE-GROUP* and represents disjunctive knowledge. (In general, a metarule takes ordinary rules as its arguments and globally controls the processing done by the rules.)

> *Rule R_{control}:*
> $\langle$control-condition$\rangle$: *NOT-NULL*(*ANYONE*(R_1, R_2))
> $\langle$hypothesis$\rangle$: NIL
> $\langle$action$\rangle$: *REMOVE-ALL*(*ANYONE*(R_1, R_2))

The $\langle$control-condition$\rangle$ part of this rule specifies that whatever one of the $\langle$action$\rangle$ parts of R_1 and R_2 is evaluated, R_{control} is evaluated. Here function *ANYONE* is defined as follows:

$$\textit{ANYONE}(R_1, R_2) = \text{IF } \textit{IS-EVALUATED}(R_1) \text{ then } R_2$$
$$\text{ELSE if } \textit{IS-EVALUATED}(R_2) \text{ then } R_1$$
$$\text{ELSE NIL}$$

Predicate *IS-EVALUATED*(R_1) returns *true* when H_1 generated by R_1 has been successfully verified: when H_1 was refuted and removed from the database, it returns *false*. Function *ANYONE* with more parameters can be defined similarly.

When activated, the ⟨action⟩ part of $R_{control}$ removes, if any, the alternative hypothesis not yet verified (H_1 or H_2) from the database and eliminates the corresponding ⟨action⟩ from the action list. This means that once either R_1 or R_2 is successfully evaluated, the reasoning by the other is forced to be terminated.

The reasoning based on the above rules is performed as follows:

1. At the Kth interpretation cycle, HG_1 generates H_1 and H_2 by applying R_1 and R_2, respectively.
2. Suppose at the Nth cycle a solution for H_1 is obtained and the ⟨action⟩ part of R_1 is evaluated.
3. At the beginning of the $(N + 1)$th cycle, ISDM evaluates $R_{control}$, which removes H_2 from the database.

Note that H_1 and H_2 cannot be simultaneously verified at the same interpretation cycle.

Metarules like $R_{control}$ can also be used to prune unpromising hypotheses when new information becomes available. That is, when a new instance is generated, it has only little information about its environments, so that it generates a group of alternative hypotheses to cope with various possible environments (i.e., cases). However, since the instance generates other hypotheses based on ordinary rules, new information may be obtained before one of the alternative hypotheses is verified. In this case, a metarule can be evaluated to eliminate some of the alternative hypotheses based on the new information. Thus, the utilization of metarules increases the flexibility of the reasoning without reducing the efficiency.

Although the above reasoning method is very useful, it may introduce serious problems into the consistency examination performed by the system:

1. Since a group of alternative hypotheses are simultaneously generated and their source instances are the same, some of them may be regarded as consistent by SLDM. That is, SLDM has no information about which group of hypotheses are mutually exclusive alternative hypotheses. If a situation formed by such alternative hypotheses is resolved, the ⟨action⟩ parts of the rules which generated them are evaluated at the same interpretation

Table 3.1. Taxonomy of Actions

Type of information processed	Example
Attribute	Fill in attributes of an instance.
	Establish symbolic relations.
Instance	Create a new instance.
	Unify instances.
Hypothesis	Remove a hypothesis.
	Update a hypothesis.
Rule (metarule)	Control the evaluation of rules.

cycle. Since metarules to manage alternative hypotheses do not take such situations into account, serious errors may be introduced.

2. To remove hypotheses from the database is an unrecoverable commitment made during the interpretation process. Thus, when one of alternative hypotheses happens to be verified by an erroneous image feature, there is no way to recover from such misrecognition. Moreover, a misrecognized instance may introduce an internal conflict into an interpretation network (see Fig. 3.8).

The current system has no facilities to prevent alternative hypotheses from being accumulated and to manage interpretation networks with internal conflicts. Therefore, much care has to be taken in writing rules based on disjunctive knowledge.

In short, although some facilities for reasoning based on disjunctive knowledge are available in SIGMA, they are not so sophisticated. In order to avoid the problems described above, we have to augment the system so that disjunctive knowledge can be described more explicitly and more complete consistency examination can be performed.

We summarize the actions discussed in this and previous sections in Table 3.1.

3.9. AN EXAMPLE OF SPATIAL REASONING

In this section, we demonstrate the process of spatial reasoning in SIGMA by using an illustrative example. The example includes a variety of reasoning activities conducted by rules: the bottom-up instantiation and the top-down hypothesis generation based on *PART-OF* relations, the

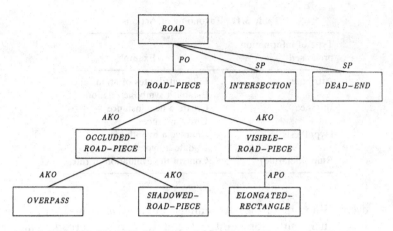

AKO: *A–KIND–OF*   PO: *PART–OF*   SP: Spatial Relation   APO: *APPEARANCE–OF*

Figure 3.13.   A world model for *ROAD*.

establishment of symbolic relations, and the unification of multiple instances. In what follows, Cycle N ($N = 1, 2, \ldots$) denotes the Nth interpretation cycle. Figure 3.13 shows the world model used in this example: although no in-conflict-with links are illustrated, object classes which are not connected by *AKO* (*A-KIND-OF*) and/or *PO* (*PART-OF*) relations are linked by in-conflict-with relations (see Section 2.4.4). (Here we use the second type of rule discussed in Section 3.6.4 to unify multiple parent instances.)

Cycle 0: Initial Segmentation

0-1. Suppose a pair of *ELONGATED-RECTANGLE* instances, ER_1 and ER_2, are extracted at the initial segmentation (Fig. 3.14a). Note that these instances are generated by LLVE.

0-2. When ER_1 and ER_2 are returned to MSE, it first generates corresponding instances of *VISIBLE-ROAD-PIECE* based on the *APPEARANCE-OF* link between *ELONGATED-RECTANGLE* and *VISIBLE-ROAD-PIECE*. Let VR_1 and VR_2 denote such instances. Specifically, in order to generate VR_1 and VR_2, MSE applies to ER_1 and ER_2 respectively a rule for instantiation stored in *ELONGATED-RECTANGLE*.

0-3. Then, MSE executes the same process based on the *A-KIND-OF* link between *ROAD-PIECE* and *VISIBLE-ROAD-PIECE*: instances of *ROAD-PIECE*, RP_1 and RP_2, are generated from VR_1 and VR_2 (Fig. 3.14b). Thus, MSE as well as ISDM applies rules to instantiate object classes at higher levels in *AKO* hierarchies.

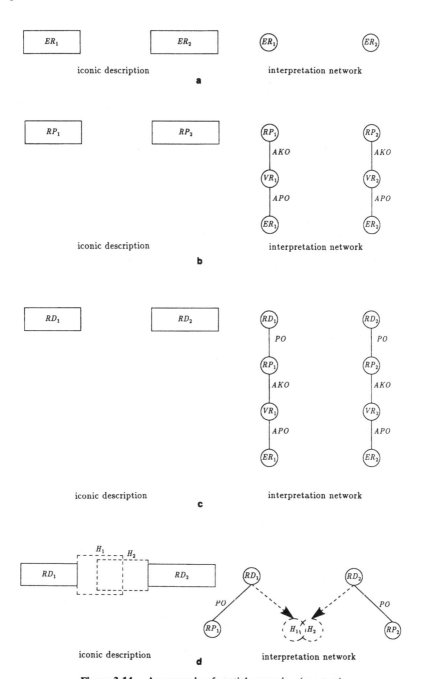

Figure 3.14. An example of spatial reasoning (see text).

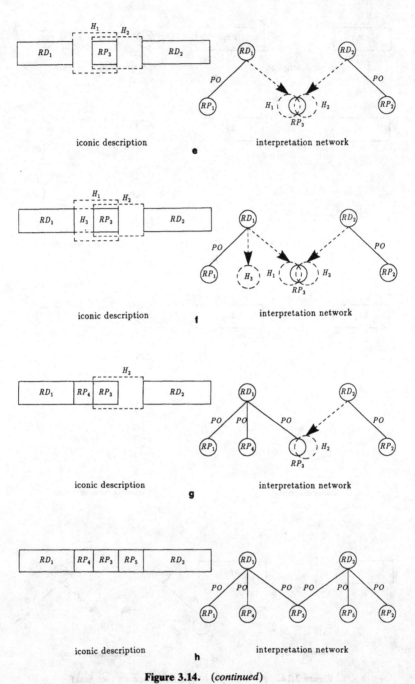

iconic description   interpretation network

e

iconic description   interpretation network

f

iconic description   interpretation network

g

iconic description   interpretation network

h

Figure 3.14. *(continued)*

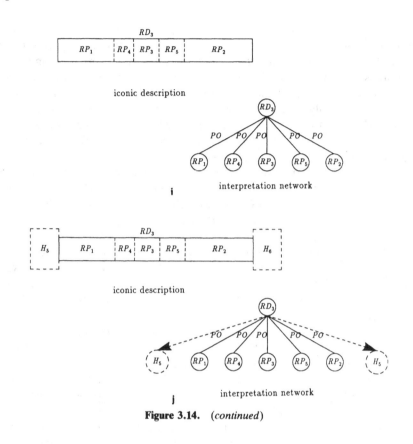

Figure 3.14. (*continued*)

Cycle 1

1-1. At the beginning of the first interpretation cycle, rules in *ROAD-PIECE* are applied to RP_1 and RP_2 to generate their parent instances of *ROAD*. This is bottom-up instantiation based on *PO* relations. Let RD_1 and RD_2 denote such instances (Fig. 3.14c).

1-2. Then, a rule in *ROAD* to generate a hypothesis for a neighboring road piece is applied to RD_1 and RD_2, respectively. By this rule application, hypotheses H_1 and H_2 are generated. Figure 3.14d illustrates this situation, where instances VR_1, VR_2, ER_1, and ER_2 are not depicted for simplicity.

1-3. Suppose H_1 and H_2 are consistent and the situation formed by them is selected by FAM.

1-4. Since no instance is participating in the selected situation, SG activates MSE to find a missing road piece. Note that the object class of the composite hypothesis generated by SG is *ROAD-PIECE*. MSE searches for an appropriate appearance of the target ROAD-PIECE using the world model in Fig. 3.13. Suppose the specialization strategy in ROAD-PIECE leads MSE to VISIBLE-ROAD-PIECE. Then, MSE transforms the composite hypothesis into a hypothesis of ELONGATED-RECTANGLE via an APO (APPEARANCE-OF) link, which in turn is given to LLVE as a goal. Suppose LLVE successfully extracts an instance of ELONGATED-RECTANGLE. Then, as was done at the initial segmentation, MSE instantiates VISIBLE-ROAD-PIECE and ROAD-PIECE based on APO and AKO links, respectively. Let RP_3 denote such an instance of ROAD-PIECE (Fig. 3.14e).

1-5. RP_3 is returned to SG, which then passes it to AS. Suppose AS first activates the ⟨action⟩ part of the rule in the action list whose cause-of-delay is H_1. Let ACT_1 denote the activated action. First ACT_1 examines the continuity between RP_1 and RP_3. (Recall that constraints on geometric relations are checked by a source instance after a solution for a hypothesis is returned.) Since in this example RP_3 is not connected to RP_1, ACT_1 suspends the establishment of the PO relation between RD_1 and RP_3, and directly activates MSE to verify a hypothesis for ROAD-PIECE which fills the gap between RP_1 and RP_3. Let H_3 denote such hypothesis (Fig. 3.14f).

1-6. Suppose MSE successfully finds an instance of ROAD-PIECE, RP_4, which satisfies H_3. Then, RP_4 is returned to ACT_1, which first establishes a PO relation between RD_1 and RP_4 since RP_1 and RP_4 are continuous. Then, the suspended PO relation between RD_1 and RP_3 is established, and the verified hypothesis H_1 is removed by AS (Fig. 3.14g). As discussed in Section 3.6.4, the unification of partially instantiated PO hierarchies can be activated at this stage in the implemented system. That is, after establishing the PO relation between RD_1 and RP_3, ACT_1 examines if RP_3 has multiple parent instances. However, since RP_3 has only one parent, RD_1, at this point, no further processing is executed by ACT_1.

1-7. After performing all this processing, ACT_1 returns control to AS. AS then activates the ⟨action⟩ part in the action list whose cause-of-delay is H_2. Let ACT_2 denote such ⟨action⟩. The same processing as above is executed by ACT_2: a new road piece instance RP_5 is detected by directly activating MSE, and a PO relation is established between RD_2 and RP_5.

1-8. When ACT_2 establishes a PO relation between RD_2 and RP_3, it finds that RP_3 has two parent instances, RD_1 and RD_2 (Fig. 3.14h). Then, ACT_2 proceeds to the unification of RD_1 and RD_2. Since RD_1 and RD_2 are consistent, a new instance of ROAD, RD_3, is generated, all relevant PO relations are linked to RD_3 (Fig. 3.14i), and the old parent instances, RD_1 and RD_2, are removed. When all this processing by ACT_2 is finished, the first interpretation cycle is completed.

Cycle 2. At the second interpretation cycle, a rule in ROAD is applied in RD_3 to generate new hypotheses, H_5 and H_6, for neighboring road pieces (Fig. 3.14j). [If we used the first type of rule discussed in Section 3.6.4 in order to unify multiple parent instances, the processing at (1-8) would be completed just after establishing a PO relation between RD_2 and RP_3 (Fig. 3.14h), and the unification would be performed at the beginning of the second interpretation cycle.]

As described above, very complicated processings can be executed during a single interpretation cycle. Moreover, very complex reasoning processes such as the direct activation of MSE and the unification of multiple parent instances can be performed by an ⟨action⟩ part of a rule.

There are several stages in the above reasoning process where the top-down analysis may fail. First, when LLVE returns *failure* to MSE, MSE performs an exhaustive search in the world model to find alternative appearance models. This search is conducted using PO and AKO links in the world model. Suppose LLVE cannot extract ELONGATED-RECTANGLE. MSE assumes that the target road piece is OCCLUDED-ROAD-PIECE and searches for its practical appearance: if appearance models of OVERPASS and SHADOWED-ROAD-PIECE are given in Fig. 3.13, MSE selects one of them and passes it to LLVE.

When all possible appearances tried by MSE fail, the failure of the top-down analysis is reported to all participating instances of the selected situation. As discussed in Section 3.8, since such *failure* itself is an informative result, the instances may perform new reasoning based on *failure*. For example, a road instance generates a hypothesis for INTERSECTION or DEAD-END when a hypothesis for ROAD-PIECE is refuted; it assumes that it cannot be extended any further because it terminates at a dead end or an intersection with another road.

Chapter 4

LLVE: Expert System for Top-Down Image Segmentation

Recently several expert systems for image processing (ESIPs) have been proposed to facilitate image analysis. They use knowledge about image processing techniques to realize effective image analysis. The Low-Level Vision Expert (LLVE) in SIGMA can be considered as one of them.

In this chapter we first provide an overview of the ESIPs proposed so far and discuss their objectives, knowledge representation, and reasoning methods. Then we describe the knowledge representation scheme and the reasoning method in LLVE, along with several experimental results. In the last section we emphasize the importance of *image analysis strategies* in realizing effective image analysis: analysis using the pyramid (multiresolution) data structure, combination of edge-based and region-based analyses, and so on. We propose two methods of representing image analysis strategies: one from a software engineering viewpoint and the other from a knowledge representation viewpoint.

4.1. EXPERT SYSTEMS FOR IMAGE PROCESSING

4.1.1. Introduction

A variety of image processing algorithms have been devised in the history of digital image processing. Although they do not work perfectly for complex natural images, their utility has been proved in various application areas, such as remote sensing, medical engineering, and office and factory automation.

In order to facilitate the wider use of digital image processing techniques, various software packages for image processing have been developed, including FORTRAN subroutine libraries (Tamu1983) and command libraries in image processing systems. As is well known, however, it is not so easy to make full use of the packages; some knowledge about image processing techniques is required to realize effective image analysis processes by combining the primitive operators in the packages.

The following are common problems encountered in designing image analysis processes:

1. *Assessment of image quality*. To assess the quality of an input image is the first step in image analysis. Although one should design an image analysis process based on this assessment, how to measure and describe image quality is a difficult problem. In particular, since the image quality often changes depending on location within the image, one needs to analyze the image thoroughly in order to make the assessment.

2. *Selection of appropriate operators*. There are many different operators (algorithms) for a specific image processing task (e.g., edge detection). They are designed based on different image models and computation schemes. One has to select an appropriate operator considering the image quality, the purpose of the image analysis, and the characteristics of the operators.

3. *Determination of optimal parameters*. Many operators have adjustable parameters; their performance is heavily dependent on the values of the parameters (e.g., threshold in binarization). How to determine optimal parameter values is another difficult problem.

4. *Combination of primitive operators*. It is often necessary to combine many primitive operators to perform a meaningful task. For example, a popular way of extracting regions from an image is to apply edge detection → edge linking → closed boundary detection. To obtain an effective combination, knowledge about image processing techniques is required.

5. *Trial-and-error experiments*. Usually it is very hard to estimate *a priori* the performance of an operator for a given image, so that one has to repeat trial-and-error experiments by modifying parameters (and sometimes operators).

6. *Evaluation of analysis result.* The process of evaluation is very important in realizing flexible image analysis: feedback analysis (Naga1984) evaluates the difference between the processing result and the desired output (i.e., the model of the object), and adjusts parameters for the analysis. Thus how to evaluate the analysis result and how to adjust parameters is an important problem in designing image analysis processes with feedback loops.

Recently, several ESIPs have been proposed to facilitate the development of image analysis processes (Nadi1984, Saka1985, Mats1986, Sued1986, Hase1987, Tori1987, Spec1988, Tamu1988). They incorporate artificial intelligence techniques—specifically knowledge about how to effectively use existing tools (i.e., image processing operators) for image analysis—to solve the above problems. The expertise stored in these systems is what we, image processing researchers, have acquired and accumulated through the development of image processing techniques.

In this section, we first discuss the significance of ESIPs in relation to both image understanding systems (IUSs) and ordinary expert systems. Then, we classify ESIPs into the following four categories and summarize their objectives, the knowledge and reasoning methods they use, and their future problems:

1. Consultation systems for image processing
2. Knowledge-based program composition systems
3. Rule-based design systems for image segmentation algorithms
4. Goal-directed image segmentation systems for object detection

4.1.2. Image Understanding Systems versus Expert Systems for Image Processing

As discussed in Chapter 1, IUSs required three types of knowledge to interpret visual scenes:

1. *Scene domain knowledge*: models of objects in the scene and their relations.
2. *Knowledge about the mapping between the scene and the image*: (physical) properties of the imaging system (e.g., focal length and viewing angle of the camera).
3. *Image domain knowledge*: types of and relations among image features such as edges, lines, and regions; how to extract them from images; and heuristics for perceptual grouping.

How to represent and use these knowledge sources is a central problem in designing IUSs.

ESIPs are different from IUSs in the following ways:

1. *Objective*: A major purpose of ESIPs is to realize effective image analysis processes by using image processing techniques (i.e., by combining primitive operators), while that of IUSs is to interpret a scene. In other words, ESIPs are developed to make full use of available image processing techniques, while IUSs are intended to realize new and versatile visual perception capabilities.

2. *Knowledge sources*: The knowledge used by ESIPs is knowledge of how to use image processing techniques as well as image domain knowledge. That is, no knowledge about the scene is used in ESIPs. On the other hand, knowledge about image processing techniques has rarely been used in IUSs.

3. *Goal specification*: "Find roads" is a typical goal given to IUSs. Since goals for IUSs are described in terms of scene domain terminology, IUSs require models of objects in the scene and knowledge about the mapping in order to establish correspondence between the object models and image features extracted from the image. On the other hand, "find rectangles" is a typical goal given to ESIPs. Since there are many possible methods to extract rectangles from an image, ESIPs require knowledge about primitive image processing operators in order to select promising ones and knowledge about image processing techniques so as to combine them effectively.

Most of the IUSs so far developed have emphasized the importance of knowledge of types (1) and (2) (scene domain knowledge and knowledge about mapping between scene and image), and many reasoning and computational methods have been developed based on such knowledge (Ball1982, Binf1982). However, we also need a large amount of knowledge to analyze image data. Knowledge of type (3) (image domain knowledge) has usually been encoded in programs, so that it is very hard to see what knowledge is used. Moreover, fixed processes of segmentation reduce the flexibility of image analysis in IUSs. For example, the poor capability of the *ribbon* detection (i.e., segmentation) in ACRONYM (Broo1981) limits its overall performance in object recognition.

Selfridge (Self1982) incorporated knowledge about image processing techniques to realize adaptive operator and parameter selection in his aerial image understanding system. The appropriate image processing operator and its optimal parameters are selected through several itera-tions of trial-and-error image segmentation. This allows flexible image segmentation and increases the reliability. Although his idea is very similar to that behind our LLVE, the knowledge for operator and parameter selection in his system is encoded in programs, and conse-quently its reasoning capability is limited.

In ESIPs we describe the image domain knowledge and the knowledge about image processing techniques explicitly (declaratively), and make it clear what knowledge is important and how we can use it effectively. By replacing a group of image analysis programs in an IUS by an ESIP, we can increase the flexibility of image analysis in the IUS. In particular, the reliability of top-down image analysis can be greatly improved by incorporating an ESIP as an image analysis module; since the model of the target image feature to be extracted and its approximate location are given in the top-down analysis, ESIPs can accurately reason about the most reliable image segmentation technique (i.e., operators and parameters) using such information (see Section 1.5 for a discussion of top-down image analysis in IUSs). This is the reason why we designed and developed LLVE as a module of SIGMA.

4.1.3. General Architecture of Expert Systems for Image Processing

Expert systems for various tasks have been developed, including signal interpretation, medical diagnosis, circuit design and troubleshoot-ing, and plant control (Haye1983). In general, the task of ESIPs is to compose effective image analysis processes based on primitive image processing operators. In this sense, ESIPs can be considered as expert design systems. The most successful system of this type would be R1 (McDe1980a), which configures computer systems suitable for a customer's requirements.

The critical differences between ESIPs and such expert design systems are:

1. Although we can describe characteristics and behaviors of elec-tronic circuits and computer hardware in terms of logical and mathematical expressions, it is very hard to describe the visual information included in an image and to specify behaviors of image processing operators.
2. Since ESIPs are given the input information in the form of raw image data, they have to analyze it to extract meaningful information.

Point (1) implies that a major objective of developing ESIPs is to formulate and describe visual information: to determine what types of image features we can extract from an image, what properties they have, and how they are related to each other. Point (2) means that ESIPs are not only expert design systems but also image analysis systems. Moreover, ESIPs should have the evaluation capability to decide whether or not the analysis result is satisfactory. Thus, ESIPs should have the capabilities of both qualitative symbolic reasoning and quantitative signal processing. Note that the integration of both qualitative and quantitative processing is also an important problem in IUSs.

Figure 4.1 illustrates the general architecture of ESIPs. It consists of the reasoning engine, knowledge about image processing techniques, a library of image processing operators, and a database recording characteristics of the input and processed image data. Operators in the library are used to analyze the image and the result is stored in the database.

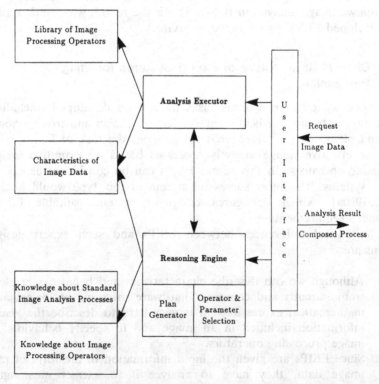

Figure 4.1. General architecture of expert systems for image processing.

The reasoning engine uses the knowledge about image processing techniques and characteristics of the image data for the reasoning.

Usually, the reasoning in ESIPs is performed at two levels:

1. *Analysis plan generation.* First, ESIPs reason about an appropriate *plan* to guide the analysis of a given image. The reasoning engine uses characteristics of the image and knowledge about standard image analysis processes to generate the plan.
2. *Operator selection and parameter adjustment.* The reasoning at this level instantiates the analysis plan into a specific image analysis process: practical operators are selected and optimal parameter values are determined. These selections are made through trial-and-error analysis of the input image. That is, ESIPs first apply promising operators and then evaluate the analysis result to replace operators and to modify parameter values.

In what follows, we provide an overview of ESIPs so far proposed.

4.1.4. Consultation System for Image Processing

A user of an interactive image processing system is usually required to select a command from a command library and to specify appropriate parameters for the command. Although several HELP facilities are available, one usually has to refer to user's manuals (on-line or written) to see the detailed usage of commands and meanings of the parameters.

A consultation system for image processing uses such manual information as its knowledge source and helps a user to select an appropriate command and parameters. Since command and parameter selection is performed under the system's guidance, the man–machine interface of the system can be greatly improved. This facility is especially useful for those with little experience in image processing.

In Sued1986 a prototype of such a consultation system, EXPLAIN, was proposed. Figure 4.2 shows the flow of the consultation. First a user specifies his purpose (e.g., enhancement, segmentation) and the approximate properties of an image to be processed (e.g., color or black and white, noise level, contrast) via a terminal. The system first reasons about a global processing plan based on both the specification given by the user and knowledge about standard image processing processes. The plan is described as an ordered sequence of *abstract image processing algorithms* such as noise elimination, edge detection, or region segmentation.

Then, the system instantiates each abstract algorithm in the sequence one by one from the beginning (Fig. 4.3): a promising practical command

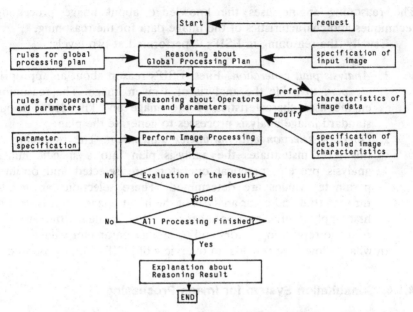

Figure 4.2. Flow of consultation (from Sued1986).

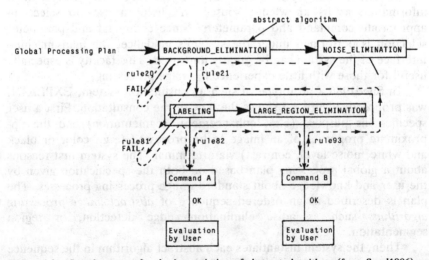

Figure 4.3. Search process for the instantiation of abstract algorithms (from Sued1986).

and its appropriate parameters are selected for each abstract algorithm. The instantiation is done through conversations on the user's detailed objective and the image quality. Conservations are guided by the system, which makes use of its knowledge about the commands.

The knowledge in this system is described by a set of production rules, which control the search process to find the appropriate command sequence. In Fig. 4.3, for example, when rule 21 for an abstract algorithm, *BACKGROUND-ELIMINATION*, is activated, it generates the process sequence *LABELING* followed by *LARGE-REGION-ELIMINATION*. Then, a rule for *LABELING* is activated to generate an executable command with appropriate parameters.

The selected command (command A in Fig. 4.3) is applied to the image and the result is immediately displayed on the monitor screen. Then the system asks the user for his evaluation. Depending on the user's evaluation, the system may replace the operator and/or modify parameters by activating other rules and retry the analysis. When no new command can be applied to satisfy the user, the system performs a large backtrack to try another global processing plan.

Besides knowledge about standard image processing processes and commands, EXPLAIN contains rules describing information about the hardware and software architectures of the image processing system, such as the number of image memories and special registers. This information is useful to hide the specific architectural features of the system from the user and enables him to think about image processing at the *logical* level.

4.1.5. Knowledge-Based Program Composition System

Many software libraries for image processing are currently available. For example, SPIDER (Tamu1983) is a FORTRAN subroutine library for image processing containing over 300 subroutines, and many image processing systems are equipped with other command libraries. The characteristics of the program modules (i.e., subroutines and commands) in a library, such as the data types for the arguments of a subroutine, are usually described in the program manual. Using such information about program modules as the knowledge source, we can develop an *automatic programming system*, which composes complex programs by combining program modules in the library. A user of the system has only to write an abstract program specification without knowing about the details of the program.

Although we usually have to improve the analysis capability and efficiency of the composed program, ESIPs of this type are useful to quickly develop application programs for image analysis.

Automatic programming has long been a dream in software engineering and artificial intelligence (Barr1982). One of the most critical problems in automatic programming is how to describe the program specification. ESIPs proposed so far use the following specification methods: (1) specification through conversation, (2) specification by abstract command, and (3) specification by example.

Although these specification methods themselves are not new, ESIPs for program composition contain the analysis executor (see Fig. 4.1), which executes (partially) composed programs so as to verify their utilities. This capability is necessary because

1. Specifications given to ESIPs are often informal and ambiguous, so that they have to repeat trial-and-error experiments.
2. Usually specifications given to ESIPs only describe image features to be extracted. Therefore, in order to verify if composed programs are satisfactory, ESIPs have to apply them to real images and to examine whether or not extracted image features satisfy the specifications.

4.1.5.1. Program Specifications through Conversation

Those systems which obtain the specification through conversation (Tori1987, Tamu1988) are very similar to the consultation system described in the previous section. All reasoning processes for plan generation and operator and parameter selection are guided by knowledge about image processing techniques and the information (i.e., specification) given interactively by the user. When the user is satisfied with the analysis result, the system outputs the program based on the analysis history stored in the system. Figure 4.4 illustrates the search tree representing the reasoning history. Each node in the tree represents a practical operator selected by the system. Some sequences of operators could not generate the satisfactory result, and the sequence illustrated by a bold line denotes a successful process. The system generates the executable program corresponding to the successful path in the tree.

Since trial-and-error analysis is required to find appropriate operators and parameters, the flexibility of ESIPs is heavily dependent on the modification process of the instantiated plan. EXPLAIN, described in Section 4.1.4, uses the ordinary tree search algorithm with backtrack to find alternatives (Fig. 4.3). DIA-Expert (Tamu1988) uses the *operation tree* to describe the image analysis process (Fig. 4.5a). The level of the operation tree is equivalent to the level of abstraction. At each level a sequence of image analysis algorithms is described. The sequence at the

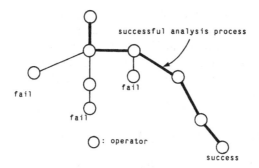

Figure 4.4. Search tree representing the reasoning history.

bottom of the tree represents the sequence of executable software modules in the program library.

In DIA-Expert all reasoning processes for plan generation, instantiation, and modification are considered as manipulations of the operation tree (Fig. 4.5b). Besides rules for these reasoning processes, the

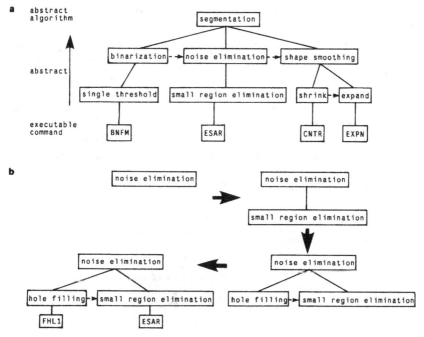

Figure 4.5. Operation tree (from Tamu1985). (a) An operation tree for segmentation. (b) Modification of the operation tree.

system contains additional rules to avoid artifacts caused by image processing operators and to enhance the effect of the operators. When these rules are activated, auxiliary nodes representing various preprocessing and enhancement techniques are added to the operation tree. These rules are useful to improve the image analysis capability of the composed program.

4.1.5.2. Program Specification by Abstract Command

When we write a program for image processing, we have to write many codes besides those for essential processing, including instructions for such steps as intialization of image data arrays and allocation of working memories. Yet when we develop a complex image analysis program using a program module library, we want to be able to devote ourselves to the function of each module without caring about such programming details.

Expert systems which generate executable programs from abstract commands greatly facilitate the development of complex image analysis programs. With such a system, we need only to specify the composition of modules in the library without knowing about their detailed syntactic and semantic structures.

Sakaue and Tamura (Saka1985) proposed an automatic program generation system using SPIDER. The system generates a complete (main) FORTRAN program from a given abstract command sequence. Figure 4.6a shows an input command sequence: (1) for an image in the standard format (SFDI), G, compute its histogram (HIST1), (2) find a threshold value from the histogram (THDS2), (3) apply binarization using the threshold (SLTH1), (4) apply connected component labeling (CLAB), (5) remove tiny regions (ERSR3), and (6) compute compactness measures for resultant regions (CRCL1). (Each command refers to the name of a subroutine in SPIDER.)

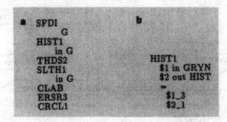

Figure 4.6. Specification by abstract commands (from Saka1986). (a) Command sequence. (b) Syntactic and semantic constraints.

The system stores the syntactic and semantic information about each argument of every subroutine in SPIDER, such as input/output discrimination, data type, and semantic usage (e.g., image data, histogram, property table). For example, Fig. 4.6b shows the information about HIST1. The first line reads "the first argument ($1) is an input argument and its data type is Gray Picture." The last three lines imply that the third attribute of the first argument (i.e., $1_3, the number of gray levels of the input picture) must be equal to the first attribute of the second argument (i.e., $2_1, the size of an array for the output histogram).

Based on such information, the system determines real arguments for each subroutine, if necessary asks a user to specify missing parameters, and generates a complete main program consisting of a set of necessary data declarations and a sequence of subroutine calls.

While specifications used by the above system describe abstract image analysis processes directly, Haas (Haas1987) proposed an automatic programming system which generates object detection programs from specified object models. Object models (i.e., specifications) are described in terms of rules of an attribute grammar. For example,

$$d1(R, S), d2(R, S) = disk(R, S)$$

$$twodisks(R, D, S) \rightarrow [d1(R, S), d2(R + D, S)]$$

describe that nonterminal twodisks consists of two terminals, d1 and d2, both of which are of disk type. A disk (i.e., d1 and d2) has two attributes, R and S, which represent its location and size, respectively. twodisks has an additional attribute D, which represent the displacement between two constituent disks.

A user gives a goal like

$$with ``twodisks": print(D, \pm2);$$

which specifies "generate a program which detects an instance of twodisks and prints D with a precision of plus or minus 2." Using prespecified object models, the system transforms the goal into an executable program:

$$d1 = diskprogram1( )$$

$$d2 = diskprogram2( )$$

$$D = d2.R-d1.R;$$

$$print(D, \pm2);$$

Here diskprogram1( ) denotes a primitive function in the program library, which analyzes an image to detect a disk and measure its properties.

4.1.5.3. Program Specification by Example

In most ESIPs a user specifies his objective, the image quality, and the evaluation result in terms of a predefined vocabulary (i.e., a set of symbolic predicates). As is well known, however, it is very hard to describe the expected analysis result and the image quality in terms of such symbolic descriptions.

In IMPRESS (Hase1987), a user can specify the goal of the analysis by a *sample figure*. That is, the request given to the system is "compose a program which extracts the sample figure from the input image."

First, the system determines a global processing plan based on the type of sample figure: point, line, or region. Then, the reversed version of each abstract algoithm in the plan is applied to the sample figure one by one from the end (see the left column in Fig. 4.7). The images generated by these reversed operations are used as references to evaluate the adequacy of operators and parameters for each abstract algorithm in the plan.

In selecting an appropriate operator and parameter (i.e., instantiation of each abstract algorithm), the system applies several promising operators with different parameters in parallel. All the processing results are compared with the reference picture which was generated from the sample figure. Then, the system selects the best operator with the optimum parameters which generated the result most similar to the reference (Fig. 4.7). The numbers associated with the pictures in Fig. 4.7 show similarity measures evaluated based on the reference picture.

Goal specification by example is natural and easy for any user. It is particularly helpful for those with no experience in image processing. Moreover, the use of sample figures is very effective to determine optimal parameters.

4.1.6. Rule-Based Design System for Image Segmentation Algorithms

Usually, algorithms for image segmentation use many heuristics to split/merge regions and lines into meaningful ones: "merge neighboring small regions with similar properties," and so on. While filtering operators for smoothing and edge detection can be designed based on well-defined mathematical models, the incorporation of heuristics into

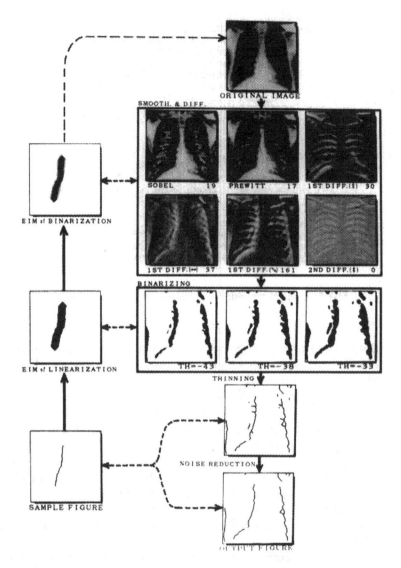

Figure 4.7. Specification by example (from Hase1987).

line and region segmentation algorithms is inevitable. Thus, to design a segmentation algorithm with high performance, we have to repeat trial-and-error experiments to test the effectiveness of the incorporated heuristics.

To facilitate such experiments a rule-based segmentation system

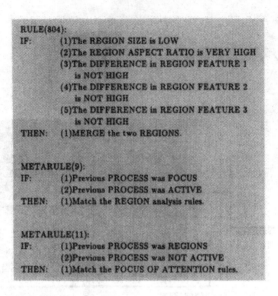

```
RULE(804):
IF:      (1)The REGION SIZE is LOW
         (2)The REGION ASPECT RATIO is VERY HIGH
         (3)The DIFFERENCE in REGION FEATURE 1
            is NOT HIGH
         (4)The DIFFERENCE in REGION FEATURE 2
            is NOT HIGH
         (5)The DIFFERENCE in REGION FEATURE 3
            is NOT HIGH
THEN:    (1)MERGE the two REGIONS.


METARULE(9):
IF:      (1)Previous PROCESS was FOCUS
         (2)Previous PROCESS was ACTIVE
THEN:    (1)Match the REGION analysis rules.


METARULE(11):
IF:      (1)Previous PROCESS was REGIONS
         (2)Previous PROCESS was NOT ACTIVE
THEN:    (1)Match the FOCUS OF ATTENTION rules.
```

Figure 4.8. Production rules for segmentation (from Nadi1984).

(Nadi1984) was proposed in which various heuristics for line and region segmentation are represented by a set of production rules (Fig. 4.8). The explicit representation of the heuristics greatly facilitates tests of their effectiveness and their modification.

The condition part of a rule is described in terms of constraints on attributes of regions and lines and their mutual spatial relations. The action part specifies a split/merge operation on regions and lines. In addition to rules for such segmentation procedures, the system stores a set of *metarules* to control the entire analysis process (Fig. 4.8). The mode of the system is switched by a metarule from region analysis to line analysis and vice versa. Using these metarules, flexible control structures can be realized, a goal which was very difficult to achieve in ordinary segmentation programs.

Although the execution time is slow in this rule-based segmentation system, its flexibility allows the rapid development of effective segmentation algorithms. Note that this expert system is different from the other ESIPs discussed in this chapter. That is, it is an expert system to develop a new image segmentation algorithm, while the others compose image analysis processes (programs) based on existing image processing operators.

4.1.7. Goal-Directed Image Segmentation System for Object Detection

This last type of ESIP is exemplified by LLVE in SIGMA. The role of LLVE was described in Section 2.1.4. In what follows we present detailed descriptions of LLVE with several experimental results.

4.2. KNOWLEDGE ORGANIZATION IN LLVE

The task of LLVE is to automatically extract image features (e.g., lines and regions) which satisfy the constraints specified in a given goal. "Find a rectangle whose area is between 100 and 200 pixels" is a typical example of the goal specification.

LLVE uses two types of knowledge to conduct automatic image segmentation:

1. Knowledge about fundamental concepts in image segmentation: types of image features extractable from an image and types of image processing operators.
2. Knowledge about image segmentation techniques: how to combine the operators effectively.

The knowledge of type (1) can be defined in a formal way, while that of type (2) involves many heuristics. In LLVE the former knowledge is represented by a network describing the *type structure* in image segmentation and the latter by a set of production rules.

4.2.1. Network Knowledge Organization

Suppose we want to extract line segments from a gray picture. Gray picture $\xrightarrow{\text{edge detection}}$ Edge Picture $\xrightarrow{\text{thresholding}}$ Edge Point $\xrightarrow{\text{linking}}$ Line Segment would be a typical analysis process to satisfy the objective. We call Gray Picture, Edge Picture, Edge Point, and Line Segment in the above example *image features*, and edge detection, thresholding, and linking *transfer processes*. That is, an image feature denotes a *type* of information extractable from raw image data, and a transfer process refers to an *abstract algorithm* which analyzes its input image feature to generate its output image feature (Fig. 4.9). Usually, to extract a specific image feature from a raw picture, we have to combine several different transfer processes as shown in the above example. We call such an ordered sequence of transfer processes a *process sequence* (Fig. 4.9).

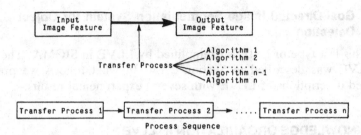

Figure 4.9. Image feature, transfer process, and process sequence.

LLVE uses image features and transfer processes as the fundamental terminology to represent knowledge about image segmentation techniques. Representing an image feature by a node and a transfer process by an arc, we can represent the knowledge by a network. Figure 4.10 illustrates the network knowledge organization LLVE, where each ellipse denotes an image feature and a directed arc a transfer process.

Usually, there are multiple different ways of extracting a specific image feature. Such different analysis processes are represented as multiple directed paths (i.e., sequences of arcs) in the network. In other words, various different process sequences are embedded into the network. For example, in Fig. 4.10 there are two major paths to extract a Region from a Gray Picture: one via Line and the other via Homogeneous Region.

4.2.2. Representing Image Features

Each image feature is represented by a *frame* (Mins1975) in LLVE. We call it an *image feature frame*. Each image feature frame consists of a set of slots representing various attributes of the image feature. For example, the image feature frame for Line contains slots representing, for example, starting point, chain code, and length (Fig. 4.11a). In addition to the slots for such attributes, every image feature frame includes two additional slots used by LLVE: a *MADE-BY* slot to represent the transfer process which generated that image feature, and a *FROM* slot to represent the *source* image feature from which that image feature was generated. The system uses these two slots to maintain the analysis history.

An image feature frame defines a *class* of image features. Real data extracted from an image are represented by its *instances*. Figure 4.11b shows an instance of image feature frame Line. All of the slots are filled with values representing the attributes of the real line, while the image

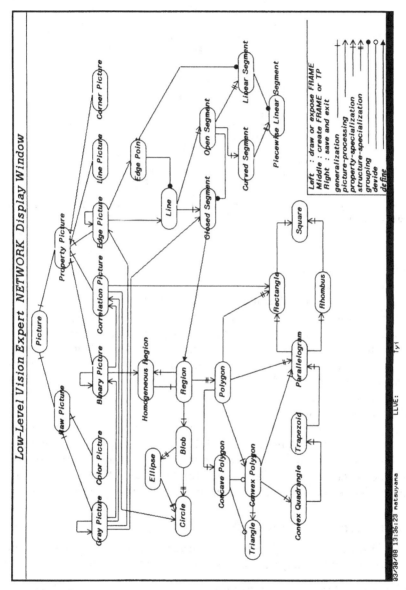

Figure 4.10. Network knowledge organization.

a

Image Feature Frame Name:	Line
Slot Name:	Starting Point
	Chain Code
	Length
	MADE-BY
	FROM

b

<Line 123>	
Slot Name	Slot Value
Image Feature Frame Name:	Line
Starting Point:	(10, 60)
Chain Code:	(2, 3, 3, 5,....)
Length:	15.5
MADE-BY:	<LINKING 890>
FROM:	<Edge Point 567>

Figure 4.11. Image feature frame. (a) Image feature frame for Line. (b) Instance of image feature frame Line.

feature frame defines the slot structure. In image segmentation, image feature frame instances are analyzed by transfer processes, and new instances are generated.*

To analyze a complex image, we usually have to apply top-down segmentation many times to the same image data. Therefore, we can use the analysis results of previously executed segmentations to perform new segmentation. LLVE includes a manager to maintain all image feature frame instances produced during the analysis (i.e., Image Feature Manager in Fig. 4.13). All requests and queries about image features are processed by this manager: generate a new image feature frame instance, find image feature frame instances which satisfy specific conditions, and so on. For example, the manager can answer a query such as "find all lines whose length is between 10 and 20." When a new goal is given to LLVE, it first asks this manager if any image feature frame instances produced so far satisfy the goal specification. If so, LLVE returns those instances as the answer without performing real image segmentation.

4.2.3. Representing Transfer Processes

A transfer process denotes an *abstract algorithm* to analyze and transform the input image feature into the output image feature. Usually

---

* The FROM slot in Fig. 4.11b is filled with only one Edge Point instance, ⟨Edge Point 567⟩, while the line consists of many points. This is because a group of Edge Point instances are represented by using a binary array to reduce the memory space; if we represented each edge point by a frame instance, a large memory space would be needed. That is, ⟨Edge Point 567⟩ represents a binary array including many edge points.

various practical algorithms can be used to realize the abstract algorithm (Fig. 4.9). Consider edge detection as an example. There are many different algorithms to realize edge detection: Sobel, Laplacian, Robert's operators, and so on.

From a knowledge representation point of view, there are several different types of transfer processes. For example, the transfer process between Edge Point and Line implies *grouping,* while the one between Parallelogram and Rectangle implies *specialization.* The types of transfer processes used in LLVE are summarized in Table 4.1. In Fig. 4.10 the type of each transfer process is illustrated by the icon attached to the head of the directed arc.

In the current implementation, no specialized reasoning is conducted depending on the types of transfer processes except for *define* type. In Fig. 4.10 Closed Segment and Region are connected by a transfer process of define type. This means that whenever a new instance of Closed Segment is produced, an instance of Region defined by it is automatically generated. This function was introduced to cope with the duality between

Table 4.1. Types of Transfer Processes

Type of transfer process	Function	Example
Generalization	Transform the specialized input image feature into the general output image feature	HOMOGENEOUS-REGION-TO-REGION
Property-specialization	Select instances of the input image feature satisfying specific conditions (the data structure is not modified)	LINE-TO-CLOSED-SEGMENT
Structure-specialization	Select instances of the input image feature satisfying specific conditions (the data structure is modified)	LINE-TO-LINEAR-SEGMENT
Picture-processing	Apply picture processing (no conceptual relation exists between the input and output image features)	GRAY-PICTURE-TO-EDGE-PICTURE
Grouping	Merge instances of the input image feature to form an instance of the output image feature	EDGE-POINT-TO-LINE
Divide	Split an instance of the input image feature to generate multiple instances of the output image feature	CONCAVE-POLYGON-TO-CONVEX-POLYGON
Define	An instance of the output image feature defined by an instance of the input image feature is generated	CLOSED-SEGMENT-TO-REGION

```
Transfer Process
Frame Name:        -----

Slot Name:         Type of Transfer Process
                   Algorithm
                   Parameters
                   Input Image Feature Frame Instance
                   Output Image Feature Frame Instance
                   Computation Time
```

Figure 4.12.   Transfer process frame.

a region (area) and a closed segment representing its boundary. That is, we consider Region, Polygon, Rectangle, and so on as representing areas, so that a closed segment (i.e., boundary-based representation) must be transformed into an area-based representation (i.e., Region) before further processing is performed.

We also use a frame to represent a transfer process. Figure 4.12 shows the structure of the frame to represent transfer processes. All transfer processes have the same slot structure, while those of image feature frames vary depending on their attributes. As in the case of image feature frames, every real analysis process performed is represented by an instance of such a frame, in which all slots are filled with real values representing the algorithm, parameters, input and output image feature frame instances, and computation time. LLVE includes a module to maintain all transfer process frame instances (i.e., Transfer Process Manager in Fig. 4.13). This manager together with Image Feature Manager maintains and answers queries about the analysis history.

In order to perform practical image segmentation, we activate transfer processes. In LLVE, Process Sequence Execution Module in Fig. 4.13 activates them. Given an input image feature frame instance, the activation of a transfer process proceeds as follows:

1. First generate a new instance of the transfer process frame to be executed. All slots of the instance are empty.
2. Store the input image feature frame instance in the corresponding slot of the created instance.
3. Select an appropriate practical algorithm to apply by using the rules for algorithm selection. (Details of the rules and rule application will be described later.)
4. Select appropriate parameters for the selected algorithm by using parameter selection rules.

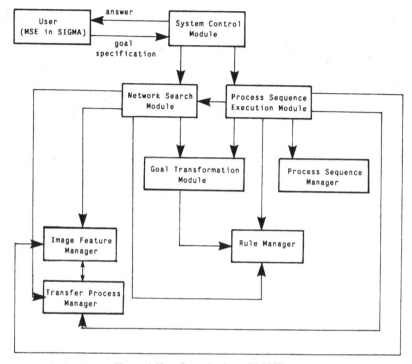

Figure 4.13. Organization of LLVE.

5. Call the image processing program (in the program library) corresponding to the selected algorithm with the selected parameters.

6. Generate a new image feature instance or instances representing the output from the program. Store it (them) in the corresponding slot in the transfer process frame instance.

7. Store the time (in seconds) of the image processing in the corresponding slot.

Thus Process Sequence Execution Module calls programs from the program library. Transfer process instances are nothing but records (data structures) to store the information about the executed programs.

4.2.4. Representing Knowledge about Image Segmentation

The network structure described above defines the fundamental terminology in image segmentation. In addition to this knowledge, a degree of practical knowledge is required to realize effective image segmentation. For example, if an image is noisy, one should smooth it before segmentation, and it is preferable to apply nonmaximum suppression to an edge picture before edge tracking. Moreover, image segmentation involves many steps of trial-and-error analysis; it is usually hard to determine the optimal algorithm and parameters before analyzing the given image. Thus we need knowledge to conduct the trial-and-error analysis: knowledge of how to select an alternative algorithm and adjust parameters based on preceding analysis results.

In LLVE, such knowledge about image segmentation is represented by production rules. Since there is a variety, we classify the rules into

Table 4.2. Types of Production Rules

Type of production rule	Function	Example

Transfer process selection rule	Select promising transfer processes during the search	AT Region
		IF ISOLATION is NO,
		THEN select
		   CLOSED-SEGMENT-TO-
		   REGION

Transfer process dependency rule	Describe the dependency between transfer processes	AT HOMOGENEOUS-
		   REGION-TO-REGION
		IF NOISE-LEVEL is not LOW,
		THEN include BINARY-PICTURE-
		   TO-BINARY-PICTURE

Cost computation rule	Estimate the processing cost of transfer processes	AT EDGE-POINT-TO-LINE
		IF algorithm equals linking and
		   COMPLEXITY is HIGH,
		THEN (equation to compute the
		   cost)

Constraint transformation rule	Transform constraints on the output image feature into those on the input image feature of the transfer process	AT REGION-TO-POLYGON FOR area IF T, THEN set POLYGON.AREA to REGION.AREA
Algorithm selection rule	Select a promising executable algorithm for the transfer process	AT EDGE-POINT- TO-LINEAR-SEGMENT IF ACCURACY is HIGH, THEN select fine Hough transform
Parameter selection rule	Select an appropriate parameter value for the algorithm	FOR thresholding IF failed before, THEN increase the threshold by 20%
Failure rule	Specify alternatives to recover from failure	AT POLYGON-TO-RECTANGLE IF T, THEN FIND-ANOTHER-PATH from Polygon

seven types. Table 4.2 summarizes the types and their functions. (The detailed function of each type of rule will be given in Section 4.3.)

Each rule has the AT part in addition to the ordinary IF and THEN parts except the parameter selection rules with the FOR part. Conceptually, the AT part specifies where in the network each rule is to be stored: at which node or arc in the network. For example, the transfer process selection rule in Table 4.2 is stored in the Region node in the network, and the transfer process dependency rule is associated with the arc between the Homogeneous Region and Region nodes. On the other hand, the parameter selection rule in Table 4.2 is associated with thresholding, a practical algorithm.

As will be described in the next section, the reasoning engine in LLVE searches the network for the most promising process sequence to extract the image feature specified in a given goal. When the search process visits a node or follows an arc, rules stored in it are activated to guide the subsequent search. In other words, image features and transfer processes define the structure to be searched and production rules control the search.

4.3. REASONING PROCESS IN LLVE

4.3.1. Configuration of the System

Figure 4.13 illustrates the configuration of LLVE. It consists of several modules for specialized tasks. The system control module is the interface between LLVE and its user (MSE in SIGMA). It controls the network search module and the process sequence execution module. The network search module searches for the most promising process sequence in the network to extract the image feature specified in the goal. The process sequence execution module (execution module, in short) executes the selected process sequence by activating the transfer processes involved in the sequence. If some transfer process fails during the execution of the sequence, the execution module reasons about what alternatives can be used to attain the goal.

All instances of image features, transfer processes, and process sequences (an instance of a process sequence is an ordered list of transfer process instances) are maintained by corresponding managers: the image feature manager, the transfer process manager, and the process sequence manager. The rule manager maintains all production rules and selects one from applicable rules. The role of the goal transformation module will be discussed later.

4.3.2. Goal Specification

When the system is started, no image feature frame instance is contained in it. So we first have to specify the original image data to be analyzed. The specified image data are represented by an image feature frame instance of either Gray Picture or Binary Picture depending on their type. (No transfer process is implemented for Color Picture.) We should also specify its attributes, such as noise level and contrast. Figure 4.14 shows an image feature frame instance representing an input gray picture.

Next a goal is given to the system to initiate the analysis. A goal specification includes (see Fig. 4.15):

1. *Type of image feature* (i.e., name of image feature frame) *to be extracted.*
2. *Local area to be analyzed.* The local area is specified by a window, an upright rectangular area defined by a pair of points corresponding to its upper left and lower right corners.
3. *Constraints on attributes of the target image feature.* The attributes correspond to the slot names of the target image feature frame. A constraint is set by specifying the minimum and maximum values for each attribute. In Fig. 4.15, for example, the area of Rectangle must be between 200 and 400 pixels.
4. *Properties of the environment in which the target image feature is embedded.* Five descriptive terms are prepared in LLVE for this purpose: *ISOLATION*, *CONTRAST*, *TEXTURE*, *NOISE-LEVEL*, and *COMPLEXITY*. *ISOLATION* is set either "YES" or NO"; this implies whether or not the target image feature is isolated in its environment. The other four can be set as either "HIGH," "INTERMEDIATE," or "LOW".
5. *Properties of the goal itself.* We can specify the properties of the goal itself in terms of *ACCURACY*, *ALLOWABLE-COST*, and *TEMPLATE*.

```
#<GRAY-PICTURE 21061632>
     MADE-BY                      : ORIGINAL
     FROM                         : NIL
     GRAY-LEVEL-ARRAY-ID          : #<ART-88-250-140 21033527>
     GRAY-LEVEL-ARRAY-TYPE        : ART-88
     OFFSET                       : (0 0)
     NOISE-LEVEL                  : LOW
     TEXTURE                      : PARTIALLY
     SCENE-TYPE                   : NATURAL
     CONTRAST                     : HIGH
```

Figure 4.14. Instance of Gray Picture representing the image under analysis.

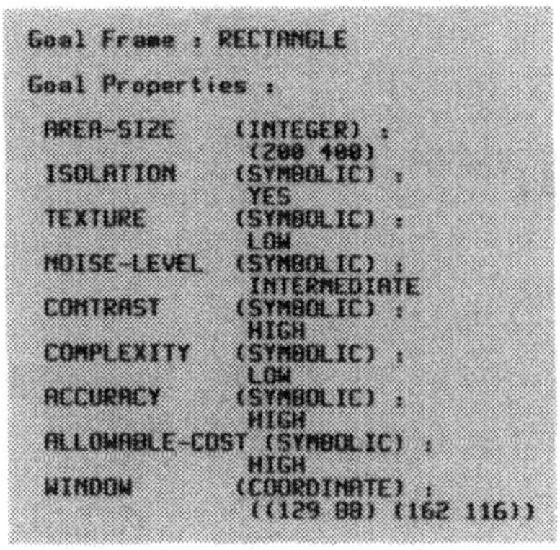

Figure 4.15. Goal specification.

ACCURACY implies how accurately the analysis should be done, and *ALLOWABLE-COST* how much computation time can be spent for the analysis. Both can be set as either "HIGH," "INTERMEDIATE," or "LOW". For example, when we have high confidence in the existence of the target image feature, *ALLOWABLE-COST* is set as "HIGH," which means "find it no matter how much computation time is required." *TEMPLATE* is set as either "YES" or "NO" to specify whether or not the analysis is to be performed by template matching using a given template. If it is set as "YES," the system asks the user to specify the template.

4.3.3. Reasoning about the Most Promising Process Sequence

Given a goal, the system first extracts the local area specified by the window from the image data under analysis, and generates a new image feature frame instance of Gray Picture (or Binary Picture) to represent the windowed image data. This instance is regarded as the original image data to be analyzed to attain the goal: all processing is performed within the window.

Then, the network search module is activated to reason about the most promising process sequence to extract the target image feature. It traverses the network guided by production rules.

4.3.3.1. Minimum Cost Search

In principle, the network search module tries to find the minimum cost process sequence connecting Gray (Binary) Picture and the target image feature in the network. The search is started from the target image

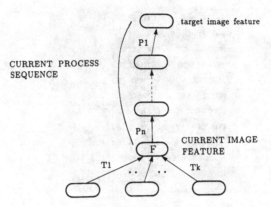

Figure 4.16. Searching in the network.

feature, and the network is traversed backward to Gray Picture. The search proceeds as follows (Fig. 4.16). [Note that the process sequence here implies a sequence of transfer processes in reverse order. Although the following algorithm is a standard best-first graph search algorithm with no heuristic function (Nils1980), we describe it in detail so as to explain the functions of the production rules later.]

Step 1. Let F and PS:(P1, P2, . . . , Pn) denote *CURRENT-IMAGE-FEATURE* and *CURRENT-PROCESS-SEQUENCE*, respectively, where Pis ($i = 1 - n$) are transfer processes and PS denotes the (backward) path connecting the target image feature and *CURRENT-IMAGE-FEATURE*. That is, the output image feature of P1 is the target image feature and the input image feature of Pn is *CURRENT-IMAGE-FEATURE*. At the beginning of the search, *CURRENT-IMAGE-FEATURE* and *CURRENT-PROCESS-SEQUENCE* are set to the target image feature and NIL [i.e., ( )], respectively. We also initialize a list named *RECORD* to NIL, which is used to store all intermediate paths (i.e., process sequences) so far searched. They are sorted according to their estimated costs: *CURRENT-PROCESS-SEQUENCE* is the minimum cost path among those stored in *RECORD*.

Step 2. Find all transfer processes whose output image feature is F. Let {T1, . . . , Tk} denote the set of such transfer processes.

Step 3. Estimate computing costs for these transfer process by applying cost estimation rules associated with them. The estimation is performed depending on the properties of the image data and the goal, so that the cost changes dynamically. This is one of the reasons why we have to search the network every time a new goal is given; if the cost could be statically computed, the minimum cost path between each pair of image features in the network could be stored in a table.

Step 4. Generate a set of new process sequences by appending each of $\{T1, \ldots, Tk\}$ to PS: $PS1:(P1, \ldots, Pn, T1), \ldots,$ $PSk:(P1, \ldots, Pn, Tk)$. Append them to and remove PS from *RECORD*. Compute the cost of each PSi $(i = 1 - k)$ by adding the cost of Ti to that of PS.

Step 5. Find the minimum cost process sequence among those in *RECORD*. Let $NPS:(NP1, \ldots, NPm)$ denote such process sequence and assign it to *CURRENT-PROCESS-SEQUENCE*. Assign the input image feature of transfer process NPm to *CURRENT-IMAGE-FEATURE*.

Step 6. If *CURRENT-IMAGE-FEATURE* equals Gray Picture or if there exists at least one instance of *CURRENT-IMAGE-FEATURE* which satisfies the constraints, then stop and return *CURRENT-PROCESS-SEQUENCE* as the answer. (See below for the constraints used at this step.) Otherwise, go to Step 2.

Note that while costs of transfer processes change dynamically for different goals and image data, they are fixed during the search. Thus, this algorithm can always find the minimum cost path in the network (refer to Nils1980 for properties of this best-first graph search algorithm).

The search is terminated before arriving at Gray Picture if an instance of some intermediate image feature frame which satisfies the constraints has already been extracted. For example, once edge detection has been applied in the window, the search stops at Edge Picture. This mechanism is useful to avoid multiple applications of the same processing to the same data.

The constraints specified in the goal are on the target image feature, while the constraints examined at Step 6 are on the features of *CURRENT-IMAGE-FEATURE*. Since the properties of the former and latter image features are different, the network search module activates the goal transformation module (Fig. 4.13) to transform the original constraints on the target image feature into those on *CURRENT-IMAGE-FEATURE*. The goal transformation module uses goal transformation rules (Table 4.2) to perform the transformation. A set of goal transformation rules are associated with each transfer process; these rules describe how to convert constraints on its output image feature into those on its input image feature. Rules of this type are applied at Step 5 after finding a new CURRENT-IMAGE-FEATURE. That is, the rules associated with NP*m* are applied to generate the constraints on the new CURRENT-IMAGE-FEATURE (i.e., the input image feature of NP*m*). (Note that the transformed constraints on the output image feature of NP*m* have been already computed.) Thus, the selected process sequence (i.e., path) is associated with sets of constraints: each set specifies required properties of each intermediate image feature (i.e., node). These constraint sets are used to evaluate intermediate analysis results produced during the execution of the selected process sequence.

4.3.3.2. Selection of Promising Transfer Processes

Usually, the computation cost alone is not sufficient to guide the search. LLVE uses the utility of transfer processes as another criterion. There are various pieces of knowledge about which of several possible analysis processes is more effective in a specific situation. For example, in case of a noisy image, apply smoothing first and use region-based analysis; many noise points would interfere with edge tracking in edge-based analysis. Such knowledge is represented by two types of production rules: those for transfer process selection and those describing the dependency between transfer processes (Table 4.2). During the search, they are activated to prune search paths and introduce auxiliary transfer processes, respectively. The pruning and introduction of transfer processes by such rules supersede the minimum cost search: an expensive path can be selected if its utility is acknowledged by rules. This is the major reason why we conduct the search every time to find the most promising process sequence; the utilities of transfer processes change very much depending on the goal and properties of the image data.

Transfer process selection rule. A set of transfer process selection rules is associated with each image feature (i.e., node in the network). They are used to select promising transfer processes from those detected at Step 2 in the above search algorithm, i.e., $\{T1, \ldots, Tk\}$. Condition parts of the rules are described in terms of the properties of the corresponding image feature and the original image data. At Step 2, if the condition of a rule is satisfied, some of $\{T1, \ldots, Tk\}$ are selected and the others are discarded. That is, the rule prunes search paths in the network: some process sequences are not expanded even if their costs are small. Rule application is performed by the rule manger, which selects and applies one from among the applicable rules.

Transfer process dependency rule. A set of transfer process dependency rules is associated with each transfer process (i.e., arc in the network). They describe the dependency of that transfer process on another transfer process. If transfer process T*i* is selected at Step 2 in the above algorithm, one of the transfer process dependency rules corresponding to it is activated if its condition is satisfied. The rule implies that before applying T*i*, another transfer process must be performed to make T*i* work effectctively and/or to reduce the artifact caused by T*i*. For example, using a rule of this type which is associated with the transfer process (Edge Linking) from Edge Point to Line (Fig. 4.10), we can represent such knowledge as "apply nonmaximum suppression first to facilitate edge point linking." Suppose that *CURRENT-IMAGE-FEATURE* is F and transfer process S1 is selected by a selection rule at Step 2 as shown in Fig. 4.17a. Suppose also that there exists a rule for S1 describing the

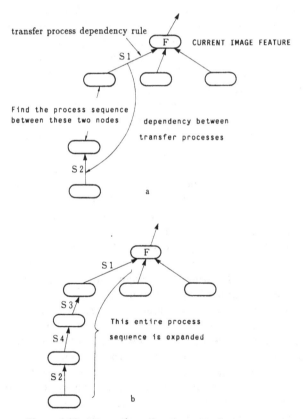

Figure 4.17. Dependency between transfer processes.

knowledge that transfer process S2 must be implemented before S1 to make S1 effective (Fig. 4.17a). In this case, the network search module calls itself recursively to find the most promising process sequence (subsequence) between the input image feature of S1 and the output image feature of S2 (Fig. 4.17b). Then, the cost of the entire subsequence detected is estimated at Step 3. At Step 4, the subsequence is added to CURRENT-PROCESS-SEQUENCE to generate a new process sequence.

Using the minimum cost search algorithm controlled by production rules, the network search module finds the process sequence with the minimum cost and the maximum utility.

4.3.4. Execution of the Selected Process Sequence

The network search module returns the selected process sequence to the system control module, which then forwards it to the process sequence execution module. Let $PS:(P1, \ldots, Pn)$ denote the selected backward process sequence. The output image feature of P1 is the target image feature specified in the goal, and the input image feature of Pn (usually Gray Picture) has an instance(s) to be processed as the source data for the analysis.

The execution module activates the transfer processes in PS one by one in the reverse order, i.e., from Pn to P1. As described before, the activation of a transfer process involves algorithm and parameter selection. A set of algorithm selection rules are associated with each transfer process and a set of parameter selection rules with each algorithm (Table 4.2). In activating the transfer process, these rules are applied to determine the appropriate executable algorithm and parameters. Conditions of algorithm and parameter selection rules are described in terms of the attributes of the input and output image feature of the corresponding transfer process, the properties of the original image data, and the history of the processing already performed. As will be described in the next section, the last type of information is used to modify algorithms and parameters when the transfer process fails to generate the desired image feature frame instances.

Instances of the output image feature are generated by activating the transfer process. The execution module examines the consistency between the attributes of the generated instances and the constraints on the output image feature which were transformed from those on the target image feature. If they are consistent, the execution module regards the transfer process as successful, and activates the next transfer process using the newly generated image feature instances as its input data. Otherwise, the transfer process is regarded as failed.

4.3.5. Recovering from Failure

As mentioned before, image processing usually involves trial-and-error analysis, so that we need knowledge about how to cope with failures of the analysis. If a transfer process fails, the execution module applies one of the failure rules associated with that transfer process. They describe what should be done to recover from failure. The execution module calls the rule manager to select one of them in case of failure.

The condition part of a failure rule is described in terms of attributes of the failed transfer process and the process history so far executed. In the action part of a failure rule, one of the following three actions can be specified:

1. *Retry.* This implies that the failed transfer process should be retried by modifying the algorithm and/or parameters (Fig. 4.18a). The new algorithm and parameters are directly specified in the failure rule.

2. *Retry-activate-from.* This rule selects one of the transfer processes executed so far in the process sequence, and requests a change in its algorithm and/or parameters. Note that the selected transfer process is usually different from the failed transfer process. In other words, a failure rule of this type considers that the cause of the failure is not in the failed transfer process but in another transfer process executed before it. After applying the failure rule, the execution module restarts the activation of the process sequence from the selected transfer process (Fig. 4.18b). That is, some of the already executed transfer processes are activated again with new data; since the selected transfer process is executed with a modified algorithm and/or parameters, it generates different output data. (An example of this failure recovery is given in the next section.) The algorithm and parameter modification is performed by the algorithm and parameter selection rules associated with the selected transfer process. For example, the parameter selection rule in Table 4.2 means that if Thresholding is requested to be retried by some failure rule, the threshold value should be increased by 20%.

3. *Find-another-path.* While actions of types (1) and (2) do not modify the process sequence under execution, this rule performs partial modification of the process sequence. Let F and G denote the input image feature of the failed transfer process and the target image feature respectively. A rule of this type requests the network search module to find another path between F and G. The execution module discards the part of the original process sequence which connects F and G, and activates the newly selected process sequence from F (Fig. 4.18c).

Usually, multiple failure rules are associated with each transfer

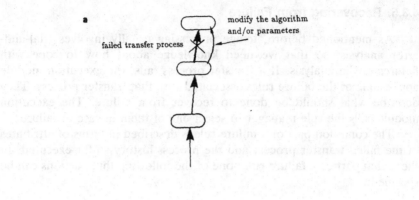

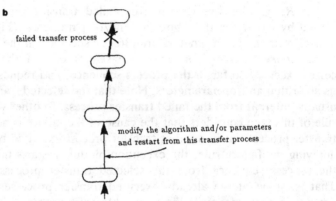

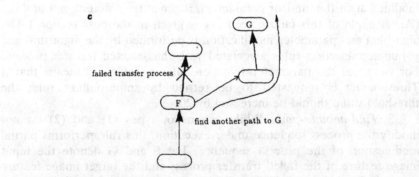

Figure 4.18. Recovering from failure. (a) RETRY. (b) RETRY-ACTIVATE-FROM. (c) FIND-ANOTHER-PATH.

process and the process sequence modified by a failure rule may fail again. Thus the trial-and-error analysis conducted by LLVE can be very complicated. Let {R1, R2, R3} and {Q1, Q2} denote sets of failure rules associated with transfer processes T1 and T2 respectively. Consider the following situation:

1. T1 fails and R1 is applied.
2. R1 is of the third type (i.e., find-another-path).
3. The new process sequence includes T2, which again fails.

In this situation, either Q1 or Q2 is applied first to manage the failure of T2. Suppose Q1 is applied. If the analysis modified by Q1 also fails, then Q2 is applied next. If the application of Q2 again results in failure, either R2 or R3 is applied, and so on. In short, the application of failure rules is done in a *depth-first* way.

If application of all available failure rules continues to result in failure, the execution module requests the network search module to find the next most promising process sequence in the network. During these nested applications of failure rules, the process sequence manager records all executed process sequences so as to avoid duplicate execution of the same process sequence.

In the following situations, the execution module stops the entire analysis process and returns "failure" to the system control module (this implies that LLVE gives up the detection of the target image feature) iff

1. All available process sequences fail, and no failure rule is left to apply.
2. The total computation time spent so far exceeds the limit specified in *ALLOWABLE-COST* in the goal. (Currently, HIGH: 300 sec, INTERMEDIATE: 200 sec, LOW: 100 sec.)

4.4. EXAMPLES OF IMAGE SEGMENTATION BY LLVE

Currently, LLVE is implemented on a Symbolics Lisp Machine and is not directly connected to MSE or GRE in SIGMA. (In the experiments described in Chapter 5, a simplified version of LLVE was used.) It supports window-based interactive graphic facilities for a (human) user to specify a goal, to monitor the reasoning and analysis processes, and to examine and edit the knowledge (i.e., network and rules). Although no accurate computation times have been measured, its response is quick enough for interactive use.

Figure 4.10 illustrates the entire set of implemented image features and transfer processes. About 400 production rules are stored and about 50 image processing operators are implemented in Lisp. Image feature frames and transfer process frames are implemented in Flavor, an object-oriented programming language in Lisp.

Figure 4.19 illustrates an example of image segmentation by LLVE. Figure 4.19a shows an aerial photograph used as test data in SIGMA. Figure 4.19b shows a goal specification given to LLVE: find a rectangle or rectangles in the image shown in Fig. 4.19a whose area is between 100 and 400 pixels and which is located in the window with [upper left and lower right corners (165, 89) and (204, 115)] respectively. COMPLEXITY refers to how complex is the environment in which the target rectangle is embedded.

Figure 4.19c illustrates the process sequence determined by the network search module. The *looping* transfer process at Binary Picture means noise elimination. Although the inclusion of such a transfer process increases the computation cost, it was incorporated to realize effective analysis by the transfer process dependency rule associated with transfer process BINARY-PICTURE-TO-HOMOGENEOUS-REGION. The reason why transfer process POLYGON-TO-RECTANGLE is selected rather than the path from Polygon to Rectangle via Parallelogram (Fig. 4.10) is that COMPLEXITY of the goal is set as HIGH; the transfer process selection rule at Rectangle is activated to prohibit the search from following the path from Rectangle to Parallelogram.

Figure 4.19d illustrates image feature frame instances generated during the execution of the selected process sequence, where transfer process POLYGON-TO-RECTANGLE fails; the extracted polygon cannot be regarded as a rectangle. Then, a failure rule associated with that transfer process is applied. Its action specifies RETRY-ACTIVATE-FROM GRAY-PICTURE-TO-BINARY-PICTURE. The heuristic represented by this rule is that since thresholding is very sensitive to the threshold value, it is reasonable to modify the threshold value if thresholding is included in the already executed (failed) process sequence. Figure 4.19e illustrates the modified process sequence, which generates the image feature frame instances shown in Fig. 4.19f. In this case, the rectangle satisfying the goal specification (i.e., $100 <$ area < 400) is successfully extracted.

Figure 4.20 illustrates another example of rectangle extraction. In this case, ISOLATION and COMPLEXITY are set as NO and LOW in the goal, respectively, while the other constraints in the goal are the same as those in Fig. 4.19. This leads the network search module to select the process sequence involving Edge Detection; the transfer process selection rule associated with Region prohibits the search from following the path

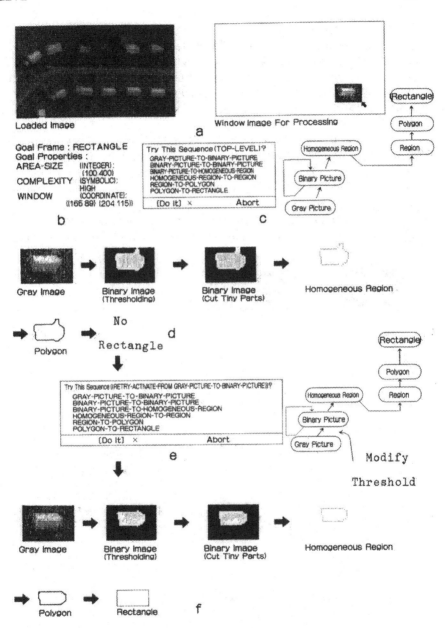

Figure 4.19. Example of image segmentation I.

connected to Homogeneous Region. The heuristic used in that rule is
that since the target rectangle is not isolated, region-based analysis will
not work well. As shown in Fig. 4.20a, however, no closed segment is
extracted by the selected process sequence (Fig. 4.20b). Then the failure
rule associated with *LINE-TO-CLOSED-SEGMENT* activates *FIND-ANOTHER-PATH*
from Line, and the new process sequence via Open Segment is generated
(Fig. 4.20b).

Figure 4.20c shows the result of executing this new process sequence.
Five instances of Closed Segement are generated by transfer process
OPEN-SEGMENT-TO-CLOSED-SEGMENT: four small closed segments and the
one corresponding to the target rectangle. Each extracted closed segment
generates an instance of Region; the transfer process between Closed
Segment and Region is of *define* type, which automatically generates an
instance of the output image feature defined by an instance of the input

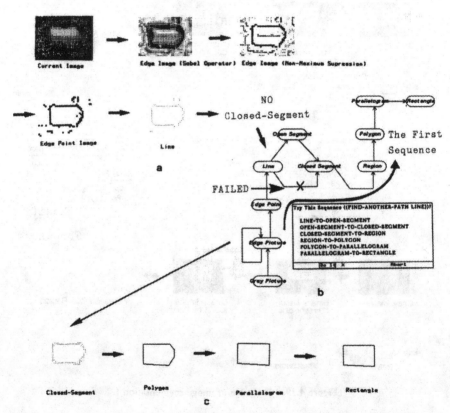

Figure 4.20.   Example of image segmentation II.

image feature. Then the process sequence execution module examines whether the properties of the generated regions satisfy the constraints propagated from the target image feature. The constraint that the area of the target rectangle must be between 100 and 400 pixels has been directly propagated to Region, which prohibits the four small regions from being used as the input of the next transfer process, *REGION-TO-POLYGON*. Thus only one instance of Region is processed in the subsequent analysis and is successfully recognized as the rectangle satisfying the goal specification.

Since all the knowledge used in LLVE is about image processing, it can analyze any image irrespective of the scene depicted in the image. Figure 4.21 illustrates the analysis of an image of an industrial part (a bolt). The first goal given to LLVE is to detect straight line segments corresponding to the head of the bolt (Fig. 4.21a–d). Three straight line segments are extracted and returned as the answer. As shown in Fig. 4.21d, however, a part of the head was not detected, so that a second request is given to LLVE to find it in a more restricted area (Fig. 4.21e).

In SIGMA or object recognition systems using LLVE, many top-down analyses of this kind are performed. The further the analysis proceeds, the more information is obtained from the image and the more accurate is the top-down segmentation that can be performed to extract missing image features. Thus LLVE is a very useful module for image understanding systems to incrementally gather information from the image.

4.5. REPRESENTING KNOWLEDGE ABOUT IMAGE ANALYSIS STRATEGIES

In the history of digital image processing, many *analysis strategies* have been proposed to increase the performance of image analysis: image analysis based on the pyramid data structure (Tani1975), a combination of edge-based and region-based analyses (Milg1979), various optimization methods (Mont1971), and so on. While these strategies are useful to improve the accuracy, reliability, and efficiency of the analysis, the expert systems for image processing proposed so far do not utilize them at all. Thus how to incorporate analysis strategies into expert systems to improve their image analysis capability is an important problem.

In Mats1988 we proposed that we can represent typical image analysis strategies by introducing three types of heterogeneous compositions of primitive image processing operators. In this section, we discuss two schemes for representing image analysis strategies: one from a

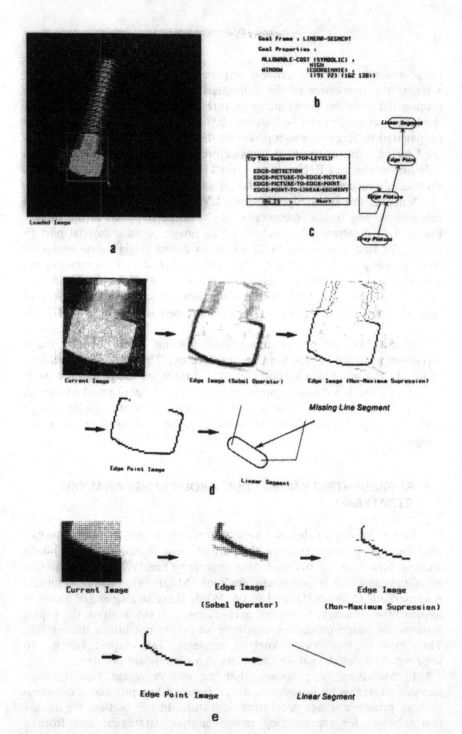

Figure 4.21. Example of image segmentation III.

software engineering viewpoint and the other from a knowledge representation viewpoint.

4.5.1. Heterogeneous Composition of Image Processing Operators

Most expert systems compose image analysis processes by combining primitive operators. However, the composition is confined to sequential combination of the operators. That is, the output of an operator is successively passed to the next operator as its input data. Although many standard image analysis processes can be composed as sequential combinations of operators, their performance is limited.

When we describe an image processing operator by a function, the result of applying O to image D is denoted by $O(D)$. The sequential composition of operators can be described as

$$O_n(O_{n-1}(\ldots O_2(O_1(D)))) \qquad (4.1)$$

where D denotes an input image and $O_1, O_2, \ldots, O_{n-1}, O_n$ are functions representing image processing operators. These functions are successively applied in this order: the innermost function is applied first to produce the data for the second function and so on. Here we omit arguments of the functions representing parameters of the operators.

We have proposed the following three types of *heterogeneous* (i.e., nonsequential) compositions to represent typical image analysis strategies and implemented them in a functional programming language for image processing: (1) composition of multiple analysis results, (2) mask-controlled operation, and (3) parameter optimization. Although we have implemented a simple interpreter for the functional programming language, we will only illustrate several programs describing complex image analysis processes without going into implementation details.

The composition of type (1) is described as

$$COMBINE(O_1(D_1), \ldots, O_n(D_n) \; by \; C) \qquad (4.2)$$

where $D_1, \ldots, D_n$ are input images for operators $O_1, \ldots, O_n$, respectively (Fig. 4.22a). C denotes a function based on which $O_i(D_i)$ are combined. Note that C is just a function which takes multiple images as its input data and combines them to produce one output image. Thus we could describe (4.2) as

$$C(O_1(D_1), \ldots, O_n(D_n))$$

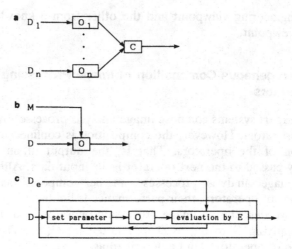

Figure 4.22. Heterogeneous compositions of operators. (a) Combination of multiple analysis results. (b) Mask-controlled operation. (c) Parameter optimization.

or more simply

$$C(D^1, \ldots, D^n)$$

where $D^i = O_i(D_i)$. In other words, function *COMBINE* is introduced to explicitly describe the combination of multiple image data, and keyword *by* to specify the function name for the combination. These syntactic structures improve the readability of programs consisting of nested function calls. (See examples given later.)

Typical examples of the processing done by C include (1) pixel-wise logical and arithmetic operations between multiple images, (2) spatial combination of local analysis results by the mask-controlled operation, and consistency examination among multiple analysis results (Milg1979). (Examples are given later.)

The mask-controlled operation is a popular method to analyze specified (focused) local areas in an image, by which focus of attention (Naga1984) in image analysis can be realized. It is described as

$$MASK(O(D) \ by \ M) \tag{4.3}$$

M is a mask image to specify local areas in image D where operator O is to be applied (Fig. 4.22b). Each pixel in the mask image has either TRUE or FALSE as its value. The output of function MASK is defined as

$$O(D(x, y)) \quad \text{if } M(x, y) = \text{TRUE}$$
$$\text{undefined} \quad \text{if } M(x, y) = \text{FALSE}$$

where $D(x, y)$ and $M(x, y)$ denote pixels in D and M respectively. Mask M itself can be an ouput of another operator which extracts *interesting* regions.

Note that we cannot write (4.3) as

$$\text{MASK}(D' \text{ by } M)$$

where $D' = O(D)$. This is because function MASK applies function O only to those areas in D that are specified by M. In this sense, function O itself is an argument of function MASK. A function which takes another function as its argument is called a *functional*. To make it explicit that MASK is a functional, we may write (4.3) as

$$\text{MASK}(O, D, M)$$

On the other hand, we use keyword *by* in (4.3) just for easy reading, as in the case of (4.2).

Parameter optimization is very useful to determine the optimal parameter value for an operator. It is described as

$$\text{OPTIMIZE}(O(D, \ldots, *, \ldots), n_1, n_2, n_3 \text{ by } D_e \text{ at } E) \qquad (4.4)$$

where D denotes an input image to be processed, O an image processing operator, D_e the reference image, and E the evaluation function. $*$ denotes the parameter for operator O to be optimized. The parameter is changed from n_1 to n_2 at the increment of n_3, and the operator O with each different parameter is applied to D. Then the results are evaluated by evaluation function E using D_e as the reference: E compares each result with D_e and measures their consistency. The function OPTIMIZE selects and outputs the best result among them: it selects the one whose consistency measure is the optimum (Fig. 4.22c). Note that function OPTIMIZE is also a functional and that keywords *by* and *at* are used for easy reading.

Using these three types of heterogeneous compositions, we can describe complex analysis processes in a compact way. For example,

$$\text{MASK}(region(D) \text{ by } binarym(\text{COMBINE}(region(D), sobel(D)$$
$$\text{by } edge\text{-}count), threshold)) \qquad (4.5)$$

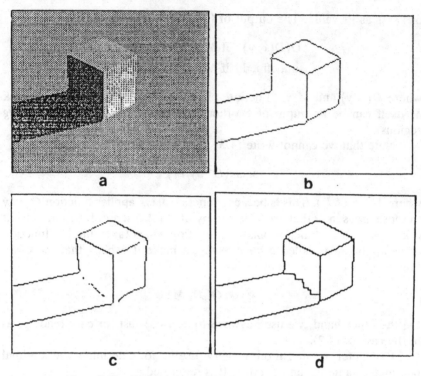

Figure 4.23. Combining edge-based and region-based analyses.

combines region-based and edge-based analyses to extract meaningful regions. Figure 4.23 shows analysis results from this program. First a pair of the innermost functions *region* and *sobel* are applied: an input image *D* (Fig. 4.23a) is segmented into regions by function *region* (Fig. 4.23b) and edges are extracted by function *sobel* (Fig. 4.23c). Note that function *sobel* calls for edge enhancement followed by binarization and produces a binary picture representing edge points. Then, these two results are combined by function *edge-count,* which enumerates the number of edge points in each region. It outputs the image where each pixel is given the number of edge points in the region to which it belongs. Function *binarym* performs binarization using the specified threshold [i.e., *threshold* in (4.5)] and outputs a logical mask image. Finally, the area specified by this mask, which denotes regions with many edge points in them, is analyzed by function *region* again to segment such non-homogeneous regions. Figure 4.23d illustrates the final result of this heterogeneous region segmentation. Note that this figure shows the regions generated by (4.5) as well as homogeneous regions which were not processed by the last mask operation. In other words, we need to apply another function *COMBINE* to obtain the picture shown in Fig. 4.23d.

Figure 4.24 shows an example of parameter optimization. It is usually difficult to determine the optimal threshold for an image of low constrast (Fig. 4.24a). Figure 4.24b shows the result of binarization by using the threshold determined by Ohtsu's method (Ohts1979).

$$OPTIMIZE(binary(D, *), 40, 60, 2 \text{ by } solbel(D) \text{ at } efunc1) \qquad (4.6)$$

realizes the binarization method proposed by Milgram (Milg1979). First function *sobel* is applied to D to detect edges (Fig. 4.24c). The evaluation function *efunc*1 examines the consistency between the resulting edge image and region boundaries obtained by the binarization operator, *binary*. Function *OPTIMIZE* changes the threshold value from 40 to 60 by 2 and outputs the binary picture processed with the optimal threshold determined by *efunc*1 (Fig. 4.24d).

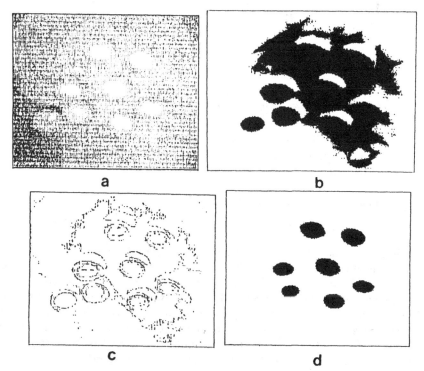

Figure 4.24. Optimization of threshold.

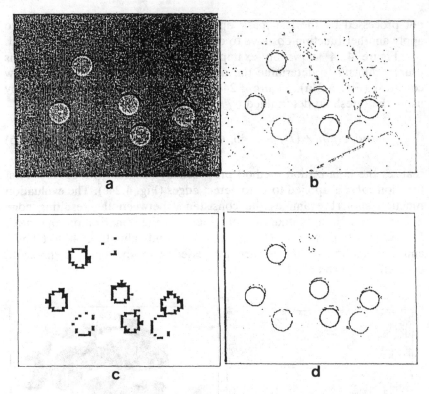

Figure 4.25.   Edge detection using multiple resolutions.

Figure 4.25 illustrates an example of edge detection using multiple resolutions. Figures 4.25a and 4.25b show an original noisy image and edge points detected from it respectively.

$MASK(binary(sobel1(D), threshold)$ *by*

$$enlarge(binarym(sobel1(shrink(D)), threshold)) \qquad (4.7)$$

realizes the plan-guided edge detection proposed by Kelly (Kell1971). First the innermost function *shrink* is applied to reduce the size of the image, and edge points are extracted from the shrunken image by functions *sobel1* and *binarym*. Function *sobel1* enhances edges and function *binarym* produces a logical mask to be used for the subsequent mask operation. Then the logical mask image representing detected edge points is enlarged, and is used as the mask for detecting edge points in the original image (Fig. 4.25c). [Functions *shrink* and *enlarge* modify the size (resolution) of an image by 0.5 and 2.0 respectively.] In the last mask

operation, functions *sobel*1 and *binary* [i.e., *binary*(*sobel*1(*D*), *threshold*]
are applied only to those pixels specified by the mask. *threshold* in (4.7)
refers to a given threshold value for binarization. Figure 4.25d shows the
image output by (4.7). By recursively applying this shrink-and-enlarge
operation, analysis based on the pyramid data structure can be realized.

As demonstrated in these examples, we can describe various image
analysis strategies by using the proposed heterogeneous compositions.
Their incorporation into ordinary image processing software will greatly
improve the performance of image analysis.

4.5.2. Representing Analysis Strategies in LLVE

Here we augment LLVE so that it can compose image analysis
processes using the analysis strategies described in the previous section.

LLVE uses a network to represent knowledge about image processing (Fig. 4.10). However, the input of each transfer process (i.e., an arc
in the network) is limited to a single image feature and the process
sequence executed is nothing but a sequential combination of transfer
processes. In order to represent the heterogeneous compositions described above, we need to incorporate transfer processes with different
input image features; COMBINE, MASK, and OPTIMIZE, described in the
previous section, can be modeled by transfer processes with multiple
input image features. We have introduced a *hypergraph* to represent such
extended transfer processes.

A hypergraph is a generalized graph (network) consisting of a set of
nodes and a set of *hyperarcs*. The generalization lies in the fact that while
an arc in a graph connects a pair of nodes, a hyperarc connects a node
with a set of nodes.

We can use a hyperarc to represent a transfer process with multiple
input image features. That is, each function for the heterogeneous
composition, COMBINE, MASK, and OPTIMIZE, is represented by a hyperarc
in the hypergraph (Fig. 4.26). In this figure, a hyperarc is illustrated by a
set of directed arcs grouped by a circular line. For example, the
binarization process in equation (4.6) can be represented by a hyperarc
connecting Gray-Picture and Edge-Point to Binary-Picture.

Image analysis using multiple spatial resolutions such as the pyramid
data structure and scale space filtering (Witk1983) has proven very
effective. However, all image features in LLVE are considered to be at a
single spatial resolution. Thus we need to extend its knowledge representation further so that image analysis based on multiple spatial
resolutions can be realized.

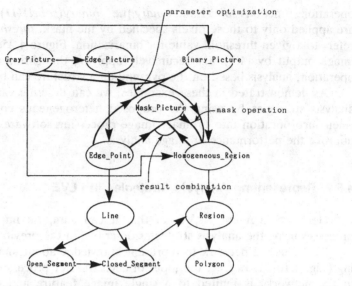

Figure 4.26. Knowledge representation by a hypergraph.

One straightforward idea is to use a hierarchical hypergraph (Fig. 4.27). At each level of this hierarchical hypergraph, a subhypergraph representing the knowledge about image features and transfer processes at a certain spatial resolution is stored. Operations which change the spatial resolution, such as shrinking and enlargement of an image, are represented by transfer processes across the levels.

However, since most image processing operators are defined irrespective of the spatial resolution, subhypergraphs at different levels

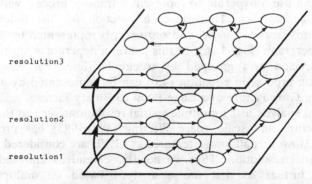

Figure 4.27. Hierarchical hypergraph.

take the same structure. In order to eliminate this redundancy, we could introduce into each image feature a parameter representing its spatial resolution. For example, Gray_Picture(s) denotes the image feature of Gray_Picture type at resolution s. Using these image features with the resolution parameter, we can compress the hierarchical hypergraph shown in Fig. 4.27 into a single layered hypergraph.

Figure 4.28a illustrates a parameterized hypergraph. Note that while the knowledge itself is represented by this single layered abstract hypergraph, it is instantiated and expanded into a multilayered hypergraph by the search process. Figure 4.28b illustrates an instantiated hypergraph with fixed resolution parameters. It represents the analysis strategy for edge detection using multiple spatial resolutions.

While the network knowledge representation in LLVE is augmented to a hypergraph with the resolution parameter, the other knowledge represented by production rules need not be changed. We can use the

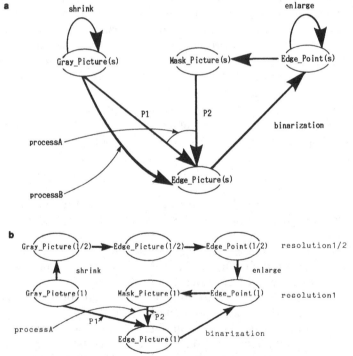

Figure 4.28. Instantiation of a parameterized hypergraph. (a) A parameterized hypergraph. (b) An instantiated hypergraph.

same search method to reason about the most promising process sequence.

For example, when we are given an input image, Gray_Picture(s) node in Fig. 4.28a is instantiated and its resolution parameter is set to 1 (Fig. 4.28b). Suppose the current *goal* is to find Edge_Point(1). Given such goal, Edge_Point(s) in Fig. 4.28a is also instantiated to Edge_Point(1). LLVE starts searching backward from the image feature node specified in the goal. The search and instantiation of Fig. 4.28a are done as follows:

1. As described above, Gray_Picture(1) and Edge_Point(1) are generated as seed nodes in the instantiated hypergraph.

2. In Fig. 4.28a, two incoming arcs (transfer processes) are attached to Edge_Point(s): *enlarge* and *binarization*.

3. Suppose a production rule associated with Edge_Point(s) prohibits the search process from following the former arc (i.e., *enlarge*); to apply transfer process *enlarge, processA* in Fig. 4.28a for the mask operation must be included in the already traversed path. In other words, transfer process *enlarge* must be used in connection with the mask operation represented by *processA* to realize plan-guided edge detection.

4. Then, the search process follows the binarization process to Edge_Picture(s). At the same time, the search process generates Edge_Picture(1) and connects it to Edge_Point(1) via *binarization* in the instantiated hypergraph.

5. There are *two* incoming arcs at Edge_Picture(s): a hyperarc representing *processA* and an ordinary arc representing *processB* (Fig. 4.28a). Suppose we take the hyperarc representing *processA*. If we followed the other arc, the search would be terminated to generate the process sequence consisting of *processB* (i.e., edge detection) and binarization.

6. Since the selected hyperarc leads to both Gray_Picture(s) and Mask_Picture(s), the search process has to continue the search from these two nodes.

7. As in (4), Mask_Picture(s) is instantiated and added to the instantiated hypergraph. On the other hand, since the input data to be processed are stored at Gray_Picture(1) [i.e., Gray_Picture(1) has been already included in the instantiated hypergraph], one of the search paths (denoted as P1 in Fig. 4.28a) is terminated at this node.

8. The other path (i.e., P2 in Fig. 4.28a) leads again to Edge_Point(s) via Mask_Picture(s). Since *processA* has been included in the already traversed path (i.e., the instantitated hypergraph so far constructed), the enlarge process is selected by the rule associated with

Edge_Point(*s*) [see (3) above]. Then a production rule associated with tranfer process *enlarge* is activated. It represents the dependency between transfer processes (see Table 4.2). The knowledge represented by this rule is that to apply transfer process *enlarge,* transfer process *shrink* must be applied first: both of these transfer processes must be included sequentially in the search path. Based on this knowledge, the sequence of transfer processes connecting Gray_Picture(1) to Edge_Point(1) via Gray_Picture(1/2), Edge_Picture(1/2), and Edge_Picture(1/2) is included in the instantiated hypergraph (Fig. 4.28b).

9. Since this search path also ends at Gray_Picture(1), the input data node, the search process is terminated. That is, the completely instantiated hypergraph has been constructed.

Although the search process seems to be complicated, the search in the hypergraph is the same as that in the AND/OR graph (Nils1980). Since many search algorithms have been developed for the AND/OR graph, we can use them for this search process.

With the augmentations described above (i.e., hypergraph knowledge representation and introduction of the scale parameter), LLVE could perform various types of heterogeneous composition of primitive operators, a modification which would substantially improve its image analysis capability. (These augmentations had not been implemented yet at the time of writing).

4.6. DISCUSSION

In this chapter we first surveyed four types of expert systems for image processing and discussed the knowledge representation and reasoning methods they use. As is seen from the references, research in this field is very active in Japan. While the system described in Tori1987 is now commercially available, most of the systems were developed to examine the feasibility of particular techniques. We believe that they are potentially useful to expand the application areas of image processing.

In Sections 4.2 and 4.3, detailed descriptions of the knowledge representation and the reasoning method in LLVE were given. As demonstrated in Section 4.4, a top-down image segmentation expert like LLVE is very effective in realizing general image understanding systems. The incorporation of such an expert allows clear separation between the scene domain knowledge and the image domain knowledge. In Section 4.5, we proposed that various image analysis strategies can be realized by using three heterogeneous compositions of primitive image processing operators. To implement these composition methods, we discussed two schemes: a functional programming language and knowledge representation using a hypergraph. By introducing image analysis strategies, expert systems for image processing can compose nontrivial, truly effective image analysis processes.

Since expert systems for image processing are very new, many problems remain to be solved: some of the problems described in Section 4.1.1 have been solved, but others are left for future research:

1. *Description of image quality and knowledge.* All systems use production rules to represent heuristics and knowledge about image processing techniques. However, the vocabularies used to write the rules are very limited. Moreover, it is very hard to describe verbally image quality, shape features, and spatial relations. As in the case of describing spatial relations (see Section 1.4), it would be useful to use fuzzy predicates to describe image quality (Naga1988). The method used in IMPRESS (Hase1987), that is, to use a sample figure for the specification, would be an interesting approach to cope with this problem.

2. *Generalization of composed image analysis processes.* Many systems compose an image analysis process (program) based on a given test image, so that operators and parameters in the composed process are selected based on specific image data. However, many users want to have a general image analysis process which works well for a set of images of similar type. So we have to generalize the composed process to make it effective for every image in the set, a process which has much to do with program generation from examples. A preliminary discussion of this problem is given in Hase1988.

3. *Evaluation of analysis results.* As noted in Section 4.1.1, the capability of evaluating analysis results is crucial in realizing effective image analysis processes. While most expert systems ask a user for evaluation, the evaluation methods themselves should be considered as important knowledge for expert systems. In the schemes described in Section 4.5, many evaluation functions are incorporated as fundamental primitives in the functional language and the hypergraph knowledge representations. We should develop new effective evaluation methods as well as new image processing operators.

4. *Translation of domain knowledge.* All knowledge used by expert systems for image processing is purely domain independent; no knowledge about a specific task domain is used. In one sense this is a big advantage; the systems can be used to analyze images in any task domain. On the other hand, when we want to develop image analysis processes for a specific task like medical examination, we have to

translate the terminology in the task domain into that in the image domain; for example, boundaries of blood vessels can be described by gray level edges in X-ray images. Usually this translation process requires much knowledge about objects in the task domain and imaging models (i.e., knowledge about the mapping between the scene and image domains). Deduction of shape from X (where X could refer to shading, texture, or contour) in computer vision (Ball1982) can be considered as an example of the (reverse) translation process based on imaging models. By introducing imaging models as new knowledge sources, the capabilities of expert systems will be greatly increased. Recall that in SIGMA MSE performs the translation between the scene and image domains.

In this context, expert systems which obtain specifications through conversation and by example have interesting characteristics. Users of the former systems usually evaluate analysis results based on their objectives in their task domains, for example, how well boundaries of blood vessels are detected. Similarly, users of the latter systems specify sample figures meaningful in their task domains, for example, sample figures that denote the boundaries of blood vessels. That is, users of these systems are working exclusively in their task domains without considering image analysis processes. Thus, such users may consider composed image analysis processes as procedural knowledge describing objects in their task domains: when the processes are applied, objects in the task domains (e.g., boundaries of blood vessels) are extracted.

Finally, some people may have doubts about expert systems for image processing themselves; since production systems and other knowledge representation schemes are just a new programming style, they believe, we cannot produce anything new from them. However, we believe that

1. Nobody would doubt that many useful application programs can be realized by using existing image processing operators. Expert systems for image processing are useful tools to facilitate the development of such application programs. They are especially helpful for those who are not specialists in image processing. In this sense, we can regard expert systems for image processing as new flexible software environments for developing image analysis programs.

2. As discussed in Section 4.1.2, how to describe and use knowledge about visual information is also an essential problem in image understanding. The central problem in expert systems for image processing is not to describe the knowledge in terms of rules but

to investigate what knowledge would be needed for image analysis. Although we do not know whether or not symbolic description is the best choice for knowledge representation, it would be useful to make explicit what we regard as knowledge, for the explicit description of knowledge facilitates the investigation of the knowledge itself.

Chapter 5

Experimental Results and Performance Evaluation

This chapter presents detailed experimental results of applying SIGMA to the analysis of high-resolution black-and-white aerial images of suburban housing developments. The objective of the analysis is to locate houses, roads, and driveways in a suburban scene.

We first present the model of suburban housing developments used in the experiments. Section 5.2 describes the Low-Level Vision Expert (LLVE) used in the experiments. This LLVE is a simplified version of the LLVE described in Chapter 4, which was separately developed at Kyoto University and is not connected to the Model Selection Expert (MSE) or the Geometric Reasoning Expert (GRE). The simplified LLVE was implemented on a VAX 780 using C, and MSE and GRE were implemented on a Symbolic Lisp Machine using Lisp and Flavor (an object-oriented language). These machines are connected by a network and exchange information via shared files.

In the latter half of the chapter, we illustrate detailed analysis results using an aerial image of a typical suburban housing development. The illustration starts with the initial segmentation process. This is followed by descriptions of how SIGMA analyzes the image in several typical situations. In the last section, we evaluate the performance of SIGMA in analyzing three different aerial images of suburban housing developments.

5.1. MODEL OF SUBURBAN HOUSING DEVELOPMENTS

Figure 5.1 shows the world model used in the experiments. Object classes are represented by nodes (boxes), and spatial relations (represented by rules) and links are represented by arcs. Object classes in the scene domain are illustrated by solid boxes: *HOUSE*, *RECTANGULAR-HOUSE*, *HOUSE-GROUP*, *DRIVEWAY*, *ROAD-PIECE*, *VISIBLE-ROAD-PIECE*, *ROAD*, and *ROAD-TERMINATOR*. Object classes in the image domain are illustrated by dashed boxes: *RECTANGLE* and *PICTURE-BOUNDARY*. They define the primitive terminology to describe object appearances as well as specify image features extractable by LLVE. Arcs are illustrated differently depending on their meanings.

Although the framework of SIGMA and its reasoning scheme are general enough to analyze three-dimensional (3D) scenes, the world

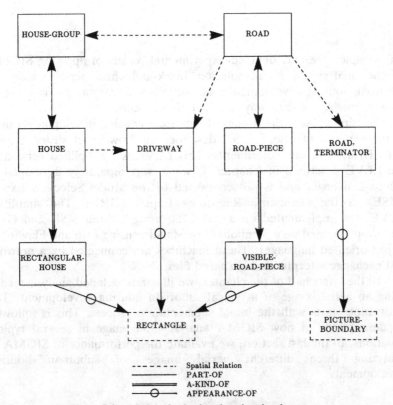

Figure 5.1. Model of suburban housing developments.

model and the implemented system apply only to two-dimensional scenes. That is, appearances of object classes in the scene domain are directly mapped to object classes in the image domain: each object class in the scene domain has at most one *APPEARANCE-OF* link. Since the major objective of the experiments is to demonstrate the utility of our reasoning scheme, we used a simple world model. [As is well known (Naga1980), we need to use 3D scene models even in analyzing aerial images, for there are many objects (e.g., shadows, bridges, and overpasses) which require 3D information for their recognition.]

There are five types of arcs between nodes: *APPEARANCE-OF* (i.e., indicating an appearance of an object class), *PO* (a composition hierarchy based on *PART-OF* relations), *AKO* (a generalization/specialization relation between a general object class and a specific object class), and spatial relation (a general spatial relation between a pair of object classes). As discussed in Section 2.4.4, in-conflict-with links are not shown for simplicity in Fig. 5.1; every pair of object classes which do not belong to the same *AKO/PO* hierarchy are connected by an in-conflict-with link. As will be discussed below, although the structure of the world model itself is simple, sophisticated control knowledge is encoded in it.

In general, when a spatial relation holds between a pair of object classes A and B, both of them can use that relation for spatial reasoning. In other words, rules for spatial reasoning are stored in both A and B, and instances of A and B use them to generate hypotheses and to establish a symbolic relation representing that spatial relation. In Fig. 5.1, such a spatial relation is illustrated by a bidirectional arc, like the dashed arc between *HOUSE-GROUP* and *ROAD*.

On the other hand, the other three arcs in Fig. 5.1 representing spatial relations are unidirectional: those from *HOUSE* to *DRIVEWAY*, from *ROAD* to *DRIVEWAY*, and from *ROAD* to *ROAD-TERMINATOR*. These undirectional arcs imply that spatial reasoning is performed only from one side (i.e., rules are stored only in one object class). For example, while a *HOUSE* instance can generate a hypothesis for *DRIVEWAY*, a *DRIVEWAY* instance cannot generate a hypothesis for *HOUSE*. The reason for this is as follows. Since the appearance of *DRIVEWAY* has few prominent features, it is very hard to correctly extract *DRIVEWAY* instances at the initial segmentation. Consequently many erroneous instances may be detected. If we allowed such instances to generate hypotheses for related objects (e.g., *HOUSE* and *ROAD*), many erroneous hypotheses would be generated, a proliferation which would increase the storage space needed and the computation time. In principle, the recognition of objects with non-prominent features should be guided by hypotheses generated from other related instances.

In this world model, a *HOUSE-GROUP* instance is defined as a group of *HOUSE* instances which are regularly arranged, have similar size and orientation, and are on the same side of a *ROAD* instance. Thus, a pair of *HOUSE-GROUP* instances are regarded as identical (i.e., they should be merged into one) if they both share a common *HOUSE* instance and are facing the same *ROAD* instance. Recall that due to the last condition (i.e., relation to *ROAD*), the situation illustrated in Fig. 3.11 can happen during the reasoning and some *PO* relations may be frozen.

A *HOUSE-GROUP* instance can generate a hypothesis about its constituent *HOUSE* based on the *PO* link between *HOUSE-GROUP* and *HOUSE*: a rule for the top-down hypothesis generation for *HOUSE* is stored in *HOUSE-GROUP*. Although only a single *PO* link is established between *HOUSE-GROUP* and *HOUSE*, each *HOUSE-GROUP* instance can be connected to many *HOUSE* instances via *PO* relations. This is because the rule for top-down hypothesis generation is repeatedly applied to a *HOUSE-GROUP* instance. That is, after finding a new constituent *HOUSE* instance satisfying a hypothesis, the rule is applied again to generate a new hypothesis so as to detect still another constituent *HOUSE* instance. Thus, the meaning of the *PO* link between *HOUSE-GROUP* and *HOUSE* is different from that of an ordinary *PO* link; the former is used to construct a composite object consisting of arbitrarily many constituent objects, while the latter connects a whole object to its specific part object. If we defined meanings of *PO* links simply by geometric transformations, as in ACRONYM (Broo1981), it would be very difficult to detect composite objects like *HOUSE-GROUP*. The same argument holds for the *PO* link between *ROAD* and *ROAD-PIECE*.

A *HOUSE-GROUP* instance can also generate a hypothesis about its facing *ROAD* using a rule (spatial relation) stored in *HOUSE-GROUP*. Once the facing *ROAD* instance is detected, its constituent *HOUSE* instances can generate hypotheses about *DRIVEWAY* based on the spatial relation between *HOUSE* and *DRIVEWAY*. In other words, in order to increase the reliability of hypotheses, the hypothesis generation for *DRIVEWAY* is suspended until a facing *ROAD* instance is found. This type of control knowledge can be easily represented in the ⟨control-condition⟩ parts of rules (see Section 3.7).

Similarly, a *ROAD* instance is defined as a group of *ROAD-PIECE* instances which have similar widths and form a continuous stretch from one *ROAD-TERMINATOR* instance to another. A *ROAD* instance repeatedly generates hypotheses about its constituent *ROAD-PIECE* instances using the *PO* link between *ROAD* and *ROAD-PIECE*: many *ROAD-PIECE* instances are grouped into a *ROAD* instance (see Fig. 3.14). Using rules in *ROAD*, a *ROAD* instance can also generate hypotheses about *ROAD-TERMINATOR*, *HOUSE-*

GROUP, and DRIVEWAY. As described above, a hypothesis about DRIVEWAY is generated after the related HOUSE-GROUP instance is detected. A ROAD-TERMINATOR hypothesis is generated when a ROAD instance cannot be extended any further (i.e., when a ROAD-PIECE hypothesis is refuted). This reasoning process is conducted based on disjunctive knowledge (see Section 3.8).

DRIVEWAY is defined as a narrow rectangular strip that connects a HOUSE instance and its facing ROAD instance. As discussed above, all arcs attached to DRIVEWAY in Fig. 5.1 are unidirectional. This implies that no rules for spatial reasoning are stored in DRIVEWAY and accordingly that its instance cannot generate any hypotheses about related objects.

ROAD-TERMINATOR in Fig. 5.1 is defined as a part of the picture boundary that intersects with a ROAD instance. In a sense, it is an artificial object which was introduced to define termination points of ROAD instances within the observed scene. As in the case of DRIVEWAY, instances of ROAD-TERMINATOR cannot generate any hypotheses; the spatial relation between ROAD and ROAD-TERMINATOR is unidirectional.

5.2. LLVE USED IN THE EXPERIMENTS

In SIGMA, LLVE is asked by MSE to extract image features (e.g., regions and lines) corresponding to the appearance of a target object in the top-down analysis. Since, as shown in Fig. 5.1, the appearances of the object classes in the scene domain are very simple, we used a simplified version of LLVE in the experiments. In what follows, LLVE stands for this simplified version.

A request given to LLVE is described by a set of linear inequalities about (constraints on) properties of a target image feature (i.e., a rectangle in the experiments). LLVE includes several different operators to perform region-based image analysis. That is, image features extractable by LLVE are confined to regions.

Given a request, LLVE selects the most appropriate analysis operator and applies it in the specified window (i.e., local area in the image). (If the window crosses over the picture boundary, it is regarded as an instance of PICTURE-BOUNDARY.) If the selected operator is able to extract at least one region, the operator returns that region as the answer. Then, LLVE approximates the returned region by a rectangle, generates an instance of RECTANGLE, and examines whether or not it satisfies the constraints specified by MSE. When the RECTANGLE instance satisfies the specified constraints, LLVE returns it to MSE. If no RECTANGLE instance satisfying the constraints can be found, the next possible operator is applied. If no applicable operators are left, LLVE returns *failure* to MSE. These image analysis processes are very similar to that used in Self1982.

Although in Fig. 5.1 the appearances of most object classes are described by RECTANGLE, different operators should be used to extract rectangles depending on their properties. In the experiments, we used four specialized operators to extract regions. They are described in the following sections.

5.2.1. Basic Image Features

A common vocabulary is needed to represent the information passed between LLVE and MSE (i.e., requests from MSE and results generated by LLVE). The vocabulary is composed of what we call *basic image features*. It is natural to select basic image features which correspond to appearances of objects in the image. For example, houses and roads in aerial images often appear as regions of homogeneous brightness. Thus we choose homogeneous regions as the basic image features.

A region is a connected component of the image. Many methods can be used to partition/segment an image into regions. Regions are used by several image understanding systems (Barr1976, Naga1980, Self1982) as the basic image features. General discussions on how various image features are represented can be found in Rose1976, Roe1982, Ball1982.

Several types of properties can be computed for a region once it is extracted. The simplest are *photometric properties* such as the maximum, minimum, mean, and standard deviation of the pixel intensity values in the region. Also in this category of properties is the average contrast of a region along its boundary: the difference between the average value of the pixels in the background which are adjacent to the region and the average value of the pixels in the region which are adjacent to the background. In our experiments, we use as the photometric properties the mean and standard deviation of pixel values in a region, and the average region contrast along its boundary.

Another type of property that can be computed is *geometric properties* such as the centroid and the area size of a region. Both are used in our experiments.

Still another type of property used in the experiments is *shape properties*. Assume the area of a region is A and the length of its boundary is P. The *compactness* of the region is defined as

$$\text{Compactness} = \frac{P \times P}{A}$$

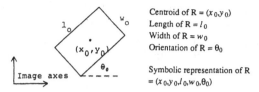

Figure 5.2.   Symbolic description of a rectangle.

Before defining additional shape properties, we first define some terms. Let R be a region. The *shrinking* operation eliminates from R the group of pixels in R which are adjacent to its background. The *width, W,* of a region is the minimum number of shrinking operators required to eliminate the entire region. The *elongatedness* of R is then defined as

$$\text{Elongatedness} = \frac{A}{W \times W}$$

The *boundary* of a region is the set of pixels in the region which are directly adjacent to its background. The boundary of a region is represented by a chain code (Free1974).

The *minimum bounding rectangle* (MBR) of a region is the smallest rectangle among those that enclose the region. MBR can be considered as the approximation of the region by a rectangle, which is the basic term to describe the appearances of objects in the scene (Fig. 5.1). To represent a rectangle symbolically, we use a 5-tuple: centroid (x and y coordinates), length, width, and orientation (Fig. 5.2). The orientation is defined as the direction of the longer side of the rectangle. Table 5.1 summarizes the region properties used in the experiments.

One special type of image feature used is those corresponding to *PICTURE-BOUNDARY*. When the window specified in a top-down request

Table 5.1.   Region Properties Used in the Experiments

Type	Property
Photometric	Mean pixel value
	Standard deviation of pixel values
	Average region contrast
Geometric	Centroid
	Area size
Shape	Compactness
	Elongatedness
	Chain code for region boundary
	Minimum bounding rectangle

crosses the boundary of the image, LLVE directly generates an instance of PICTURE-BOUNDARY and returns it as the answer. PICTURE-BOUNDARY instances are represented in the same way as those of RECTANGLE: they are also described by rectangular regions (without photometric properties).

5.2.2. Thresholding Operators

Thresholding is a popular operator for region extraction. This operator assumes that an image can be partitioned into *object* and *background*, and divides pixels into either the object or the background based on their intensity values. Two varieties of this operator were used in the experiments.

The first operator is the upper threshold operator. It is described as follows: (Window: a local area to be analyzed, x: pixel, $p(x)$: value of pixel x):

Step 1: Set *lower-bound* = $\min\limits_{x \in \text{Window}} p(x)$, *upper-bound* = $\max\limits_{x \in \text{Window}} p(x)$, and set *threshold* = *lower-bound*.

Step 2: For every x in Window, mark x as a part of the object if $p(x) >$ *threshold*. Otherwise mark it as a part of the background.

Step 3: Construct connected regions of the object and compute their properties.

Step 4: Collect all regions whose properties satisfy the given constraints. If the collection is nonempty, return it and exit.

Step 5: Increment *threshold* by 1. If *threshold* exceeds *upper-bound*, report *failure* and exit. Otherwise go to step 2.

Another operator is the lower threshold operator:

Step 1: Set *lower-bound* = $\min\limits_{x \in \text{Window}} p(x)$, *upper-bound* = $\max\limits_{x \in \text{Window}} p(x)$, and set *threshold* = *upper-bound*.

Step 2: For every x in Window, mark x as a part of the object if $p(x) <$ *threshold*. Otherwise mark it as a part of the background.

Step 3: Construct connected regions of the object and compute their properties.

Step 4: Collect all regions whose properties satisfy the given constraints. If the collection is nonempty, return it and exit.

Step 5: Decrement *threshold* by 1. If *threshold* is less than *lower-bound*, report *failure* and exit. Otherwise go to step 2.

LLVE could make use of the description of the target image feature given by MSE in order to constrain *upper-bound* and *lower-bound* more precisely. For example, suppose the goal description provided by MSE specifies that the intensity of the object appearance ranges between 18 and 35. LLVE should use this constraint to set *lower-bound* and *upper-bound* for the upper-threshold operator as

$$lower\text{-}bound = \max\{18, \min_{x \in \text{Window}} p(x)\}$$

$$upper\text{-}bound = \min\{35, \max_{x \in \text{Window}} p(x)\}$$

5.2.3. Blob Finder

Regions with small compactness values are called *blobs*. We implemented a specialized operator to detect blobs. We first segment an image into homogeneous regions and select compact ones as blobs based on their compactness values.

To segment an image into homogeneous regions, we first convolve the image with a Laplacian operator. The places where the convolved image changes sign (i.e., positive–negative) correspond to edges in the original image (Hild1979). If we assume that the regions to be extracted are small and lighter than the background, they correspond to those areas with positive values in the convolved image. Thus, by applying the upper threshold operator with *threshold* = 0, the convolved image is transformed to a binary image representing light homogeneous regions.

The Laplacian operator used computes the difference of average intensity values in two cocentric square masks of different sizes. The Laplacian's scale is specified by the sizes of the two masks. A uniform weight is assigned to every point in each mask.

In order to reduce the effect of noise, we applied a large-scale Laplacian operator to the image. However, this often causes artifacts in the convolved image. For example, two separated regions in the image may be merged into a connective positive region in the convolved image. To recover from such artifacts, we apply an 8-connected shrinking operation after binarizing the convolved image. This can break a merged region into several smaller regions at narrow bottleneck areas. All newly generated regions as well as the original ones are regarded as possible candidate regions for blobs. Finally, compact regions are selected from the candidate regions and are returned as the output.

5.2.4. Ribbon Finder

A *ribbon* is defined as a region whose elongatedness value is high and whose width is constant. As discussed above, we can extract homogeneous regions by the Laplacian operator. Then, we decompose the extracted regions into subregions to extract elongated ones with constant width. The ribbon finder performs the processing as follows:

1. We first apply a topology-preserving 8-connected thinning operation (Rose1976) to the binarized convolved image. This operation produces the skeleton map of the homogeneous regions.

2. A branch point is a point on a skeleton which is adjacent to at least three different skeleton points. After computing the skeleton of each region, we delete all branch points on the skeleton. For each separated skeleton segment that results, we compute its ideal width as follows:

Step 1: Compute the histogram of the width of the region along a skeleton segment. For a point P on a skeleton segment, we define the width of the region at P, $width_P$, as the average distance from the background to the points in the 3×3 neighborhood of P. The distance from the background can be computed by applying the distance transform (Rose1976) to the binarized convolved image.

Step 2: The width which is the mode of the histogram is regarded as the ideal width of the skeleton segment.

3. Let w be the ideal width of a skeleton segment. A point P on the segment is regarded as being on the skeleton of a ribbon iff $w - 1 \leq width_P \leq w + 1$. We call points satisfying this condition ribbon skeleton points. By grouping 8-connected ribbon skeleton points, axes of ribbons are constructed. The width of a ribbon is defined as the ideal width of the original skeleton segment.

4. The *expansion* operator expands a region R by one pixel: include in R the set of pixels in the background which are adjacent to R. To compute the ribbon from its axis, we first apply the expansion operator to the axis the same number of times as the width of the ribbon, and then intersect the resultant region with the original homogeneous region.

5. By applying (2)–(4) for each homogeneous region, we have a group of candidate regions. Finally, elongated regions are selected and are returned as the output.

5.2.5. Operator Scheduling and Selection

LLVE uses a select-and-schedule strategy to extract requested image features: first select all applicable operators and schedule the order of their application. In order to implement this strategy, we used a decision table and defined the priority order among the segmentation opeators.

With each segmentation operator, we associate a selection criterion using linear inequalities involving the region properties defined in Table 5.1. These inequalities indicate when each operator can be chosen to satisfy the given request. Table 5.2 is the decision table used in the experiments.

We define the priority (from high to low) among the operators as follows:

Ribbon finder
Blob finder
Upper threshold operator
Lower threshold operator

In order to select operators, we first combine the constraints included in a request given to LLVE with the selection criterion of each operator. If the combined constraints are solvable, we say the selection criterion is satisfied and regard that operator as applicable. For example, suppose the following constraints are included in the request:

$$R \text{ is a rectangle}$$

$$\text{Compactness}(R) < 17$$

$$200 < \text{area of}(R) < 250$$

$$R \text{ is in Window X}$$

LLVE selects the blob finder, the upper threshold operator, and the lower threshold operator, since the selection criterion of the ribbon finder is not satisfied.

Table 5.2. The Decision Table Used in the Experiments

Selection criterion	Operator
Region-contrast < 0	Upper threshold operator
Region-contrast > 0	Lower threshold operator
Elongatedness ≤ 10	Blob finder
Compactness ≥ 18	Ribbon finder

All selected operators are put into a queue. Then, LLVE orders the operators in the queue in descending priority sequence. Finally, LLVE applies each operator in the queue one by one until the request is satisfied (i.e., at least one region satisfying the constraints is returned by some operator) or no more operators are left in the queue. In the latter case, LLVE returns *failure* to MSE.

5.3. INITIAL SEGMENTATION

At the very beginning of the analysis, there are no object instances or hypotheses in the Iconic/Symbolic Database. Thus, MSE analyzes the world model and determines object classes with simple appearances. Then it asks LLVE to extract a set of image features corresponding to their appearances. Once these image features are extracted, MSE generates corresponding object instances and inserts them into the Iconic/Symbolic Database. This is the initial segmentation process in SIGMA to detect *seed* object instances, from which the analysis of the entire scene is initiated.

In what follows, we will use the aerial image shown in Fig. 5.3 to illustrate various intermediate analysis results. The size of the image is 248 × 140 and the intensity is in the range of 0 to 63.

5.3.1. Initial Segmentation Goals

The appearance of each object class can be generated by MSE using the world model. In the experiments, however, a list of image feature descriptions (called the I-set) is provided directly to MSE as the initial segmentation goals.

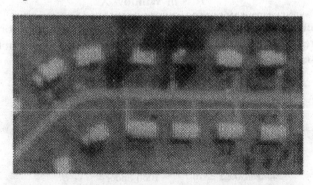

Figure 5.3. An aerial image of a suburban housing development.

We want to locate houses and roads in the image first, because their appearances have prominent features. Their appearances can be described by compact rectangles and elongated rectangles, respectively, and are usually lighter than the background. Thus, the following descriptions are used as the I-set of the initial segmentation process:

/* extract light compact rectangles */
hypothesis H_{blob}:
    target object = RECTANGLE,
    in-window = whole image,
    rectangle.elongatedness $\leq$ 10,
    rectangle.compactness < 18,
    rectangle.region-contrast > 3,
    180 < rectangle.area-of < 360

/* extract light elongated rectangles */
hypothesis H_{ribbon}:
    target object = RECTANGLE,
    in-window = whole image,
    8 < rectangle.width < 20
    rectangle.elongatedness > 10,
    rectangle.length > 10,
    rectangle.compactness $\geq$ 18,
    rectangle.region-contrast > 3

5.3.2. Verifying Initial Goals

Given H_{blob}, LLVE selected the following segmentation operators in descending order of priority:

Blob finder
Upper threshold operator

The ribbon finder and the lower threshold operator were not selected since their selection criteria were not satisfied.

The blob finder successfully extracted regions from the image in Fig. 5.3. After their MBRs were computed, they were described symbolically by rectangles: RECTANGLE instances were generated. Figure 5.4 illustrates the RECTANGLE instances detected by the blob finder. Then, properties of the RECTANGLE instances were compared with H_{blob}. All nine rectangles in Fig. 5.4 satisfied H_{blob} and were returned to MSE. Note that since the result computed by the blob finder was not empty, the upper threshold

Figure 5.4. *RECTANGLE* instances detected by the blob finder.

operator was not applied. Based on the returned *RECTANGLE* instances, MSE instantiated *RECTANGULAR-HOUSE* and inserted the generated instances into the Iconic/Symbolic Database.

Next, MSE provided LLVE with H_{ribbon}. For this task, LLVE selected the following segmentation operators in descending order:

Ribbon finder
Upper threshold operator

This time, the blob finder and the lower threshold operator were not selected. Figure 5.5 illustrates the *RECTANGLE* instances extracted by the ribbon finder. (As was the case above, the upper threshold operator was not applied.) Among these 14 *RECTANGLE* instances, 10 instances satisfied H_{ribbon}; the lowest and the upper three small ones did not satisfy the constraints. Based on each of the *RECTANGLE* instances satisfying H_{ribbon}, MSE generated an instance of *ROAD-PIECE*. Note that the two lower

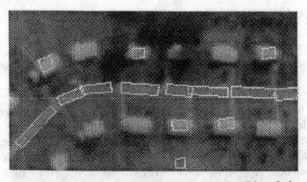

Figure 5.5. *RECTANGLE* instances detected by the ribbon finder.

RECTANGLE instances in Fig. 5.5 (corresponding to houses) were mis-recognized as *ROAD-PIECE* instances. At this stage of the analysis, even such incorrect object instances are regarded as possible interpretations. (Recall that SIGMA does not make any decision about which possible interpretations are correct during the analysis process.)

It should be noted that no instances of *DRIVEWAY* and *ROAD-TERMINATOR* are generated at the initial segmentation even if their appearances are defined in the world model (Fig. 5.1). This is because their appearances are not sufficiently prominent to extract corresponding image features by uniform image segmentation operators; we need contextual information to guide the segmentation.

5.4. CONSTRUCTING PARTIAL INTERPRETATIONS

A *situation* is classified into one of the following cases based on how the Solution Generator (SG) computes its solution:

Case 1: SG discovers an existing object instance in the Iconic/Symbolic Database which satisfies the given composite hypothesis, and all source object instances which generated the hypotheses are satisfied with the proposed solution (i.e., discovered object instance). In this case, symbolic relations between the source object instances and the discovered object instance are established.

Case 2: SG discovers an existing object instance in the Iconic/Symbolic Database which satisfies the given composite hypothesis. But some of the source object instances are not satisfied with the proposed solution. In this case, the initial hypotheses by such source object instances are modified based on the solution discovered by SG.

Case 3: SG cannot find any object instance in the Iconic/Symbolic Database which satisfies the given composite hypothesis. In this case, the top-down segmentation process is performed.

In the following sections, we illustrate these cases using several examples. In these examples, we use solid rectangles and circles to denote object instances and dashed ones to denote hypotheses.

5.4.1. Case 1—Discovering an Existing Object Instance

Consider the situation shown in Fig. 5.6 (not all hypotheses in Fig. 5.6b are depicted in Fig. 5.6a). Table 5.3 describes the database entities

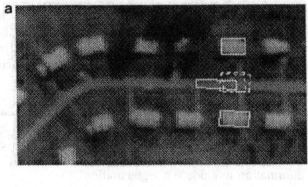

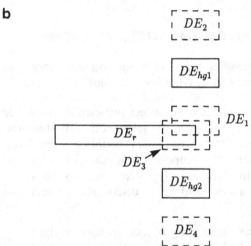

Figure 5.6. Situation including an object instance I. (a) Iconic description. (b) Graphic description.

Table 5.3. Descriptions of the DEs in Fig. 5.6b

DEs	Type	Generated by
DE_r	*ROAD* instance	
DE_{hg1}	*HOUSE-GROUP* instance	
DE_{hg2}	*HOUSE-GROUP* instance	
DE_1	*ROAD* hypothesis	DE_{hg1}
DE_2	*ROAD* hypothesis	DE_{hg1}
DE_3	*ROAD* hypothesis	DE_{hg2}
DE_4	*ROAD* hypothesis	DE_{hg2}

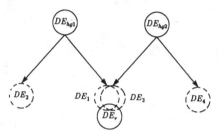

Figure 5.7. Interpretation network related to the situation in Fig. 5.6.

DEs, (instances and hypotheses) related to the situation. Figure 5.7 shows the portion of the interpretation network related to this situation.

When the Focus of Attention Mechanism (FAM) selects the situation S_1 whose P-set is $\{DE_1, DE_3, DE_r\}$, SG proposes DE_r as the solution to the composite hypothesis constructed for this situation. The Action Scheduler (AS) activates those actions in the action list whose causes-of-delay are either DE_1 or DE_3. Figure 5.8 shows the interpretation network after the actions are executed. By resolving the situation S_1, spatial relations between ROAD instance DE_r and HOUSE-GROUP instances DE_{hg1} and DE_{hg2} are established. It should be noted that two symbolic relations are established by resolving a single situation.

Note that hypotheses DE_2 and DE_4 are removed after resolving the situation S_1. This is done by a metarule stored in HOUSE-GROUP. As discussed in Section 3.8, it realizes reasoning based on disjunctive knowledge. That is, although each HOUSE-GROUP instance can be along only one ROAD instance, two alternative hypotheses about ROAD are generated at first; no information is available about on which side of a HOUSE-GROUP instance its related ROAD instance is located. Then, once one of the alternative hypotheses is verified, the metarule is activated to remove the other hypothesis. In general, the removal of hypotheses implies a commitment (decision) made by the system during the analysis, which cannot be reversed later.

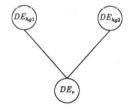

Figure 5.8. Interpretation network after resolving the situation in Fig. 5.6.

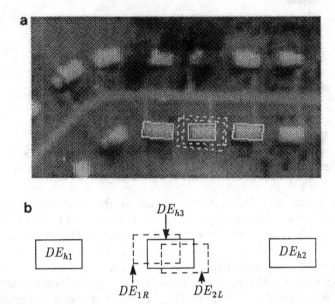

Figure 5.9. Situation including an object instance II. (a) Iconic description. (b) Graphic description.

Consider another situation shown in Fig. 5.9. The DEs related to this situation are described in Table 5.4. (*HOUSE-GROUP* instances are not depicted in Fig. 5.9b for simplicity.) Figure 5.10 shows the portion of the interpretation network related to this situation.

When FAM selects the situation S_2 whose P-set is $\{DE_{1R}, DE_{2L}, DE_{h3}\}$, SG proposes DE_{h3} as the solution of the composite hypothesis constructed for this situation. First, AS returns DE_{h3} to DE_{hg1}. Since the returned *HOUSE* instance DE_{h3} is similar to its constituent *HOUSE* instance DE_{h1}, DE_{hg1} accepts the solution. However, since DE_{h3} already belongs to the *HOUSE-GROUP* instance DE_{hg3}, DE_{hg1}

Table 5.4. Descriptions of the DEs in Fig. 5.9b

DEs	Type	Generated by
DE_{hg1}	*HOUSE-GROUP* instance	
DE_{h1}	*HOUSE* instance	
DE_{hg2}	*HOUSE-GROUP* instance	
DE_{h2}	*HOUSE* instance	
DE_{hg3}	*HOUSE-GROUP* instance	
DE_{h3}	*HOUSE* instance	
DE_{1R}	*HOUSE* hypothesis	DE_{hg1}
DE_{2L}	*HOUSE* hypothesis	DE_{hg2}

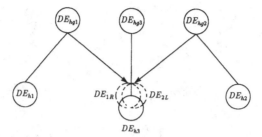

Figure 5.10. Interpretation network related to the situation in Fig. 5.9.

and DE_{hg3} are merged into a new *HOUSE-GROUP* instance, DE_{hg4} (Fig. 5.11a). Next, AS returns DE_{h3} to DE_{hg2}. Similarly, DE_{hg2} accepts the solution and is merged with DE_{hg4} to generate DE_{hg5} (Fig. 5.11b).

By resolving the situation S_2, three *HOUSE-GROUP* instances are merged into one single *HOUSE-GROUP* instance. As discussed in Section 3.6.4, these merging operations are executed by evaluating the ⟨action⟩

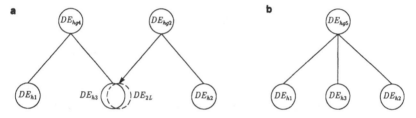

Figure 5.11. Interpretation network after resolving the situation in Fig. 5.9. (a) Interpretation network after merging DE_{hg1} and DE_{hg3}. (b) Interpretation network after merging DE_{hg2} and DE_{hg4}.

Figure 5.12. Merged *HOUSE-GROUP* instance.

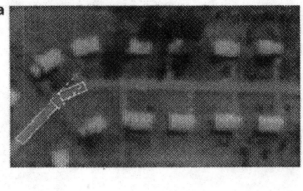

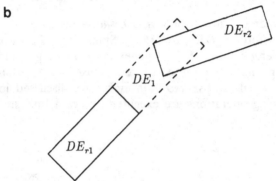

Figure 5.13. Situation including an object instance III. (a) Iconic description. (b) Graphic description.

part of the rule in *HOUSE-GROUP*: that for the top-down hypothesis generation based on the *PO* link between *HOUSE-GROUP* and *HOUSE*. Figure 5.12 shows the resulting *HOUSE-GROUP* instance. (This rectangular region was depicted by the Question and Answer Module based on the network in Fig. 5.11b.) Note that the merging operation is also an unrecoverable commitment made by the system.

5.4.2. Case 2—Modifying Hypotheses

Some source object instance which generated a hypothesis may not be completely satisfied with the solution proposed by SG. In this case, the original hypothesis is modified by the source instance depending on the solution proposed. We illustrate such a case with two examples.

Figure 5.14. Interpretation network related to the situation in Fig. 5.13.

Table 5.5.   Descriptions of the DEs in Fig. 5.13

DEs	Type	Generated by
DE_{r1}	ROAD instance	
DE_{r2}	ROAD instance	
DE_1	ROAD-PIECE instance	DE_{r1}

Consider the situation shown in Fig. 5.13. Table 5.5 describes the related DEs and Fig. 5.14 illustrates the portion of the related interpretation network.

When FAM selects the situation S_3 whose P-set is $\{DE_1, DE_{r2}\}$, SG proposes DE_{r2} as the solution to the hypothesis DE_1. That is, DE_{r2} is returned to DE_{r1} as its potential right adjacent road piece. However, since DE_{r2} is not adjacent to DE_{r1}, DE_{r1} records the solution DE_{r2} internally and retracts the hypothesis DE_1 at this interpretation cycle. Then, in the next interpretation cycle, using the description of DE_{r2}, DE_{r1} generates a new hypothesis DE_2 which fills the gap between DE_{r1} and DE_{r2}. Figure 5.15a shows the initial hypothesis DE_1 while Fig. 5.15b

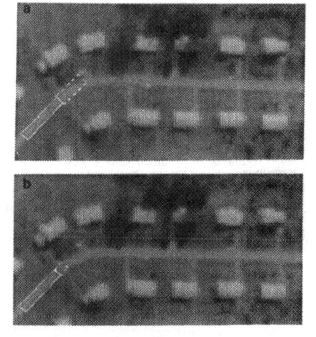

Figure 5.15.   Modification of a hypothesis. (a) Initial hypothesis. (b) Modified hypothesis.

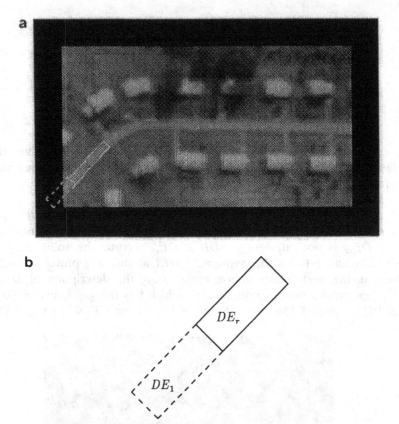

Figure 5.16. Situation which directs the failure-driven analysis. (a) Iconic description. (b) Graphic description.

shows the modified hypothesis DE_2. In short, by resolving the situation S_3, the initial hypothesis is modified and a new and more accurate hypothesis is generated.

Consider another situation shown in Fig. 5.16. Table 5.6 describes the related DEs and Fig. 5.17 illustrates the related interpretation network. When FAM selects the situation whose P-set is $\{DE_1\}$, the top-down analysis is activated. However, since DE_1 is out of the picture

Table 5.6. Descriptions of the DEs in Fig. 5.16

DEs	Type	Generated by
DE_r	*ROAD* object	
DE_1	*ROAD-PIECE* hypothesis	DE_r

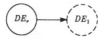

Figure 5.17. Interpretation network related to the situation in Fig. 5.16.

boundary, SG reports to DE_r that DE_1 is refuted because it is out of the picture boundary. Then, DE_r removes the hypothesis DE_1 and records the cause of failure. In the next interpretation cycle, DE_r generates a hypothesis about ROAD-TERMINATOR (Fig. 5.18). Thus, through refuted hypotheses, object instances are able to modify initial hypotheses. In other words, object instances can learn about their surrounding environments by analyzing the causes of the failures of refuted hypotheses. This *failure-driven* analysis is very effective in analyzing complex scenes; since the same class of objects are often embedded in various environments, they first investigate the environments by generating tentative hypotheses and then switch them to appropriate ones depending on the environments. As discussed in Section 3.6.4, such failure-driven analysis is also conducted based on disjunctive knowledge.

5.4.3. Case 3—Directing the Segmentation

When no object instance is included in a selected situation, the top-down analysis is activated by SG. We illustrate this process in the following two examples.

During the analysis of the image in Fig. 5.3, the following composite hypothesis, say CH_a, was given to SG:

target object = HOUSE,
in-window: W_1,

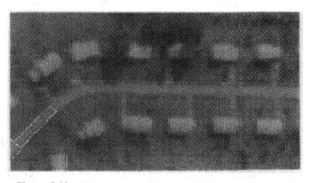

Figure 5.18. Newly generated ROAD-TERMINATOR hypothesis.

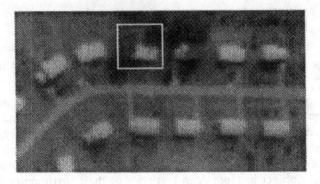

Figure 5.19. A window analyzed by LLVE.

> house.elongatedness ≤ 10,
> house.compactness < 18,
> 275 < house.area-of < 325,

The window W_1 is shown in Fig. 5.19.

SG passed CH_a to MSE, which then used the world model to determine a proper appearance of *HOUSE*. In our implementation, MSE translated CH_a to the following description and gave it to LLVE:

> target object = *RECTANGLE*,
> in window: W_1,
> rectangle.elongatedness ≤ 10,
> rectangle.compactness < 18,
> 275 < rectangle.area-of < 325

LLVE first activated the blob finder, which failed to extract any region. Then LLVE activated the upper threshold operator, which successfully extracted a region by setting the threshold value at 24. LLVE generated the *RECTANGLE* instance and returned it to MSE, which then instantiated *HOUSE* (via *RECTANGULAR-HOUSE*) and returned it to SG (Fig. 5.20).

Figure 5.21 shows another window where the same top-down analysis was performed. Note that since the target house is partially covered by a tree, its appearance differs much from that of the others. In this case LLVE applied the blob finder, the upper threshold operator, and the lower threshold operator, but could not extract any region which satisfied the given constraints. LLVE returned *failure* to MSE. Since there were no alternative appearances to try, MSE also returned *failure* to SG. If more sophisticated knowledge about *HOUSE* appearances were

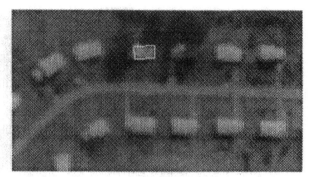

Figure 5.20. *HOUSE* instance generated by SG.

available, MSE could use such knowledge to modify the request to LLVE
and would activate it again.

Figure 5.22 shows an interesting example of finding a *DRIVEWAY*
instance by top-down analysis. Since driveways are usually narrow and
have low contrast, it is quite difficult to detect them without detailed
contextual information (e.g., location, orientation, length, and width). In
our world model, a *HOUSE* instance can generate a *DRIVEWAY* hypothesis
only after its facing *ROAD* instance is detected. Similarly, a *ROAD* instance
can generate a *DRIVEWAY* hypothesis only after it finds a facing *HOUSE-
GROUP* instance. These conditions are introduced to generate accurate
DRIVEWAY hypotheses. Figure 5.23 shows the portion of the interpretation
network related to the situation in Fig. 5.22: DE_h is linked to DE_{hg} via a
PART-OF relation and a spatial relation between DE_{hg} and DE_r has been
established. Table 5.7 describes the DEs involved in this situation.

When FAM selects the situation whose P-set is $\{DE_1, DE_2\}$, SG
activates MSE to search for a *DRIVEWAY* instance. MSE examines the

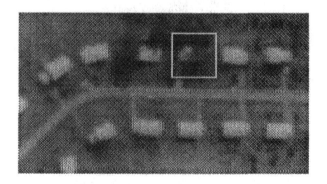

Figure 5.21.   Another window analyzed by LLVE.

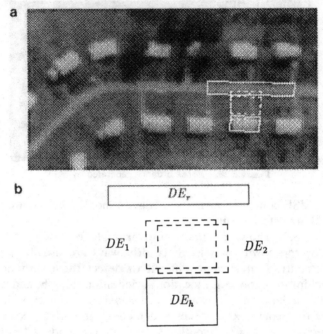

Figure 5.22. Situation which activates the top-down analysis. (a) Iconic description. (b) Graphic description.

world model to find the appearance of *DRIVEWAY*. In this case, the following request was generated and passed to LLVE:

> target object = *RECTANGLE*,
> in window: W_2,
> rectangle.elongatedness ≥ 4,

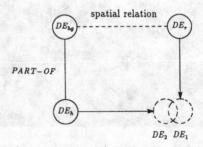

Figure 5.23. Interpretation network related to the situation in Fig. 5.22.

Table 5.7. Descriptions of the DEs in Fig. 5.22

DEs	Type	Generated by
DE_r	ROAD instance	
DE_h	HOUSE instance	
DE_{hg}	HOUSE-GROUP instance	
DE_1	DRIVEWAY hypothesis	DE_r
DE_2	DRIVEWAY hypothesis	DE_h

$$\text{rectangle.compactness} > 25,$$
$$\text{rectangle.orientation} = 90°$$

The window W_2 is shown in Fig. 5.24.

Figure 5.25 shows the RECTANGLE instance extracted by the upper threshold operator of LLVE. As this example demonstrates, provided with informative goals, even simple image segmentation operators such as thresholding can detect nonprominent image features. As will be illustrated in the next section, most driveways were correctly extracted by the top-down goal-directed segmentation.

5.5. PERFORMANCE EVALUATION

In this section, we evaluate the performance of SIGMA in analyzing three different aerial images.

The analysis by SIGMA terminates when all the hypotheses created have been verified or refuted. Figure 5.26 illustrates the final interpretation neworks for the aerial image shown in Fig. 5.3. The iconic descriptions of all recognized object instances are shown in Fig. 5.27.

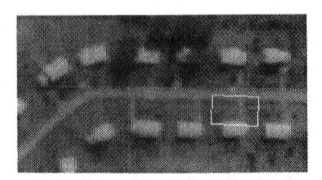

Figure 5.24. Window analyzed by LLVE to resolve the situation in Fig. 5.22.

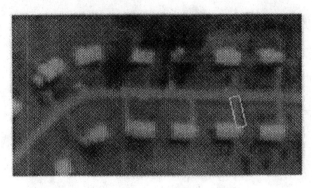

Figure 5.25. *RECTANGLE* instance generated by LLVE.

As discussed in Section 3.1.4, we select the largest interpretation network and eliminate all networks in-conflict-with it. We regard the largest network and remaining networks as the final interpretation. By this process, some incorrect object instances are detected and eliminated. In Fig. 5.26, two small networks which are in-conflict-with the largest one are eliminated. They were constructed from two *RECTANGLE* instances (ribbons) in Fig. 5.5 which were initially misrecognized as *VISIBLE-ROAD-PIECE* instances (Fig. 5.27g).

Even after this process of eliminating conflicting networks, there may still remain incorrect object instances; if interpretation networks constructed from incorrect instances do not cause any conflict with the largest network, the incorrect instances are included in the final inter-pretation. For example, if the lowest *RECTANGLE* instance in Fig. 5.5 were recognized as a *VISIBLE-ROAD-PIECE* instance during the initial segmenta-tion process, the network constructed from it would be in-conflict-with no other networks.

Table 5.8 summarizes the analysis results of Fig. 5.3. The initial and final numbers of instances of each object class and the numbers of correct and incorrect instances are listed. Note that the statistics for *VISIBLE-ROAD-PIECE* and *RECTANGULAR-HOUSE* and *PICTURE-BOUNDARY* are the same as those for *ROAD-PIECE*, *HOUSE*, and *ROAD-TERMINATOR*, respectively. Ini-tially, 23 *RECTANGLE* instances were extracted by the initial segmentation process (Figs. 5.4 and 5.5), among which 9 were interpreted as *HOUSE* instances and 10 as *ROAD-PIECE* instances, 14 more *RECTANGLE* instances were extracted by the top-down image segmentation during the inter-pretation process. All the *DRIVEWAY* instances were detected by the top-down analysis. These statistics well demonstrate the effectiveness of the top-down analysis.

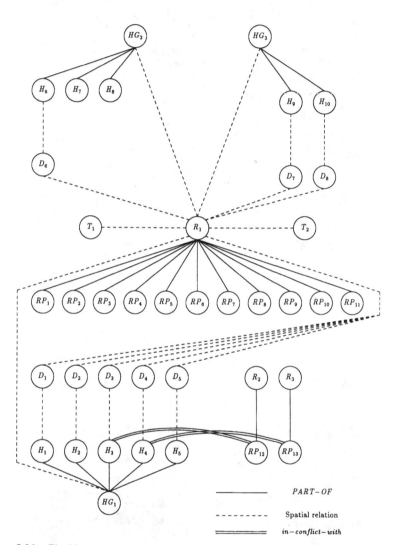

Figure 5.26. Final interpretation networks of Fig. 5.3. R: *ROAD* instance; RP: *ROAD-PIECE* instance; HG: *HOUSE-GROUP* instance; H: *HOUSE* instance; D: *DRIVEWAY* instance; RT: *ROAD-TERMINATOR* instance. (Instances of *RECTANGULAR-HOUSE*, *VISIBLE-ROAD-PIECE*, *RECTANGLE*, and *PICTURE-BOUNDARY* are not shown.)

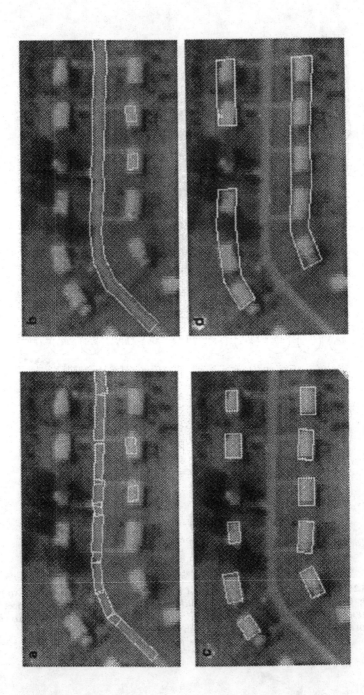

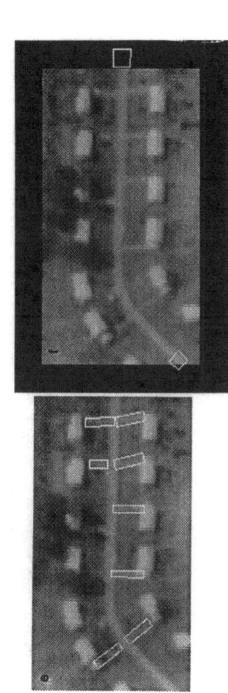

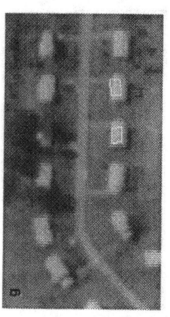

Figure 5.27. Iconic descriptions of recognized object instances in Fig. 5.3. (a) *ROAD-PIECE* instances. (b) *ROAD* instances. (c) *HOUSE* instances. (d) *HOUSE-GROUP* instances. (e) *DRIVEWAY* instances. (f) *ROAD-TERMINATOR* instances. (g) Incorrect *ROAD* instances.

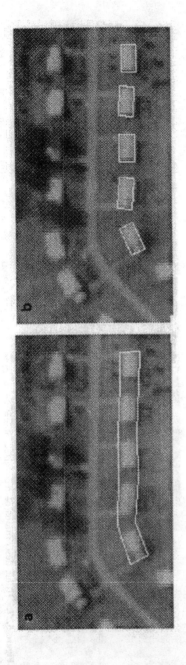

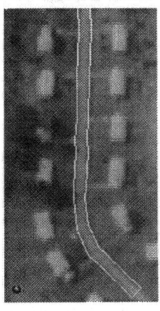

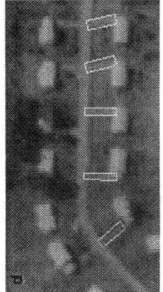

Figure 5.28. Information retrieval by using the Question and Answer Module. (a) A *HOUSE-GROUP* instance with more than four *HOUSE* instances. (b) Constituent *HOUSE* instances. (c) *ROAD* instance related to the *HOUSE-GROUP* instance in (a). (d) *DRIVEWAY* instances related to the *HOUSE-GROUP* instance in (a).

Table 5.8. Statistics of the Analysis Result of Fig. 5.3

Object class	Initial No.	Final No.	Correct No.	Incorrect No.
RECTANGLE	23	37		
HOUSE	9	10	10	0
HOUSE-GROUP	0	3	3	0
ROAD-PIECE	10	13	11	2
ROAD	0	3	1	2
DRIVEWAY	0	8	8	0
ROAD-TERMINATOR	0	2	2	0

Figure 5.28 illustrates some results of the information retrieval by the Question and Answer Module (QAM). Figure 5.28a shows a HOUSE-GROUP instance with more than four HOUSE instances. Figure 5.28b shows its constituent HOUSE instances, Fig. 5.28c the facing ROAD instance, and Fig. 5.28d the DRIVEWAY instances connecting the constituent HOUSE instances to the facing ROAD instance. QAM retrieves such information by traversing the interpretation networks in Fig. 5.27 and depicts the retrieved information iconically.

One HOUSE and three DRIVEWAY instances were not detected (see Figs. 5.27c and e); although the hypotheses about them were generated and the corresponding windows were analyzed by MSE and LLVE, no object instance was recognized. The reasons for this are: (1) the deficiency of the LLVE used in this experiment, (2) the simplicity of the world model, and (3) the insufficiency of the input information.

In order to check (1) above, we analyzed the windows by using the LLVE described in Chapter 4. Figure 5.29 shows the analysis results. First, the window in Fig. 5.29a was analyzed. This window is almost the same as the one shown in Fig. 5.21. In this experiment, we could successfully extract the rectangle shown in Fig. 5.29c. However, since the house is partially occluded, the size of the extracted rectangle is too small to be recognized as a HOUSE instance. Thus, we need to augment the world model so that such occluded houses can be recognized.

On the other hand, although the LLVE in Chapter 4 tried to extract a rectangle in the window shown in Fig. 5.29d, it could not detect any rectangle even if it applied all available segmentation methods. Figure 5.29f shows the polygon which was extracted during the segmentation. It was not regarded as a rectangle because of two positions. Considering the image data in the window (Fig. 5.29e), however, this polygon should have been recognized as a DRIVEWAY instance. That is, in order to recognize this driveway, we need to introduce another appearance model of DRIVEWAY into the world model.

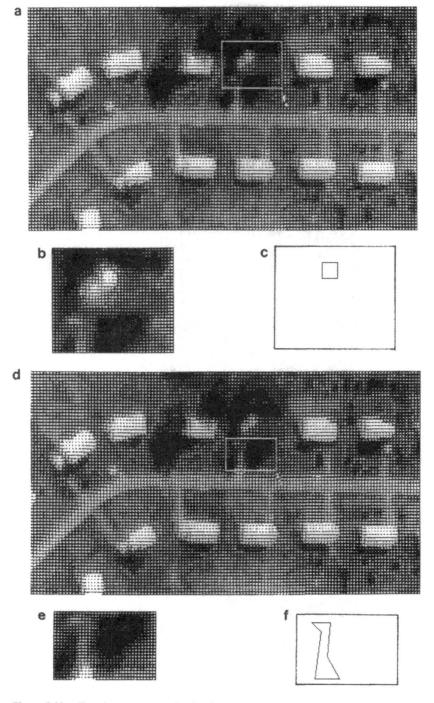

Figure 5.29. Top-down segmentation by the LLVE described in Chapter 4. (a) Window I for rectangle detection. (b) Image in window I. (c) Extracted rectangle. (d) Window II for rectangle detection. (e) Image in window II. (f) Extracted polygon.

Figure 5.30. Second test aerial image.

Although two windows including the two other missing driveways were analyzed by the LLVE, no meaningful image features could be detected. This is because the contrast of these driveways is very low, and it is almost impossible to detect them from this black-and-white image. We need more informative input image data, like color or multispectral images.

Similar experiments were performed using two other aerial images. Figure 5.30 shows a 435 × 150 aerial image. The constructed interpreta-

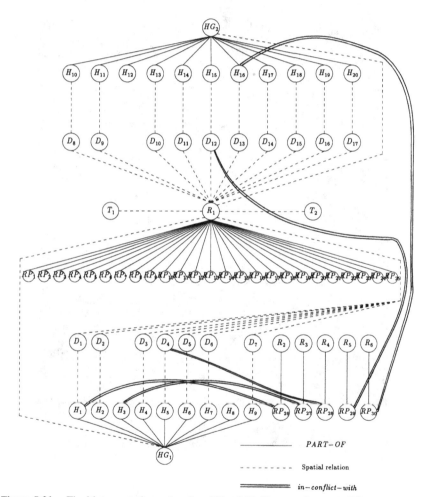

Figure 5.31. Final interpretation networks of Fig. 5.30. R: *ROAD* instance; RP: *ROAD-PIECE* instance; HG: *HOUSE-GROUP* instance; H: *HOUSE* instance; D: *DRIVEWAY* instance; RT: *ROAD-TERMINATOR* instance. (Instances of *RECTANGULAR-HOUSE*, *VISIBLE-ROAD-PIECE*, *RECTANGLE*, and *PICTURE-BOUNDARY* are not shown.)

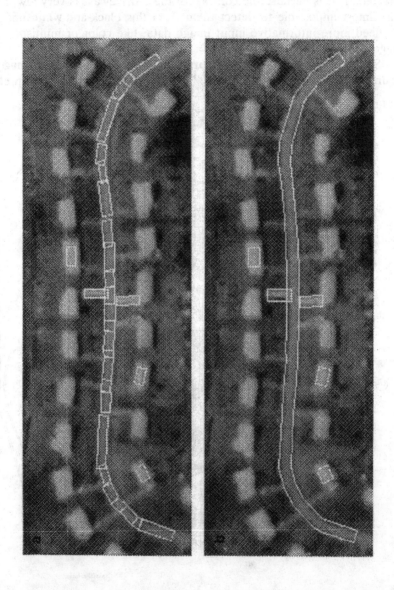

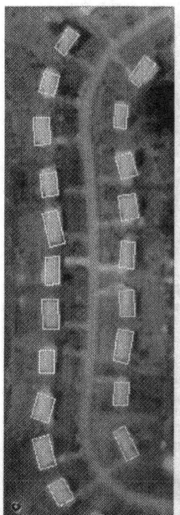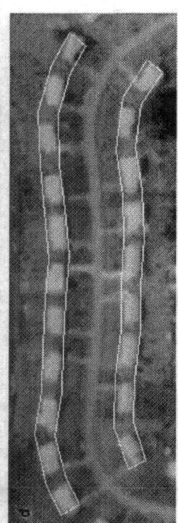

Figure 5.32. Iconic descriptions of recognized object instances in Fig. 5.30. (a) *ROAD-PIECE* instances. (b) *ROAD* instances. (c) *HOUSE* instances. (d) *HOUSE-GROUP* instances. (e) *DRIVEWAY* instances. (f) *ROAD-TERMINATOR* instances.

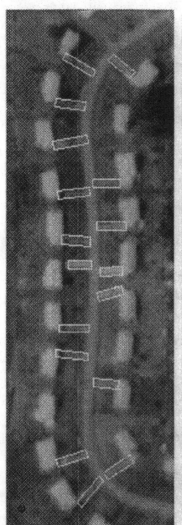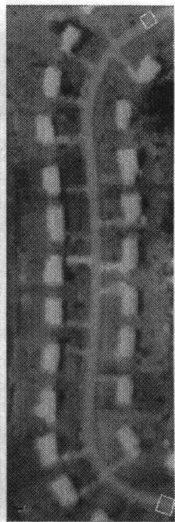

Figure 5.32. (continued)

Table 5.9. Statistics of the Analysis Result of Fig. 5.30

Object class	Initial No.	Final No.	Correct No.	Incorrect No.
RECTANGLE	43	69		
HOUSE	18	20	20	0
HOUSE-GROUP	0	2	2	0
ROAD-PIECE	19	30	25	5
ROAD	0	6	1	5
DRIVEWAY	0	17	17	0
ROAD-TERMINATOR	0	2	2	0

tion networks are shown in Fig. 5.31. The iconic descriptions of all recognized object instances are shown in Fig. 5.32.

Table 5.9 summarizes the statistics of the analysis results of Fig. 5.30. The initial segmentation extracted 43 RECTANGLE instances, among which 18 were recognized as HOUSE instances and 19 as ROAD-PIECE instances, 26 more RECTANGLE instances were extracted by the top-down image segmentation. Two RECTANGLE instances which were initially recognized as ROAD-PIECE instances were also recognized as DRIVEWAY instances (see Figs. 5.32a and e). In-conflict-with relations were established between these instances (Fig. 5.31). In this example, all houses were successfully detected, while three driveways were not detected. Again, although the hypotheses about such missing objects were generated and examined by MSE and LLVE during the interpretation process, they could not be detected mainly due to the low contrast of the image data. Although there were five incorrect ROAD instances, all the interpretation networks including them are in-conflict-with the largest (correct) network (Fig. 5.31).

Figure 5.33 is a 305 × 140 aerial image. Its interpretation networks are shown in Fig. 5.34, and the iconic descriptions of all recognized

Figure 5.33. Third test aerial image.

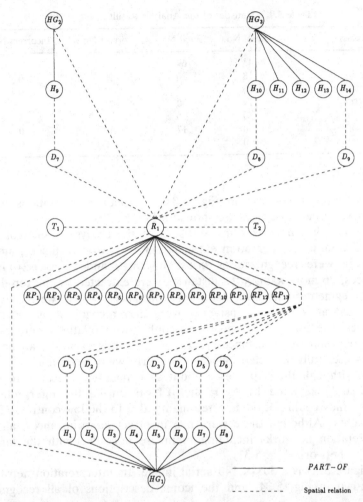

Figure 5.34. Final interpretation networks of Fig. 5.33. R: *ROAD* instance; RP: *ROAD-PIECE* instance; HG: *HOUSE-GROUP* instance; H: *HOUSE* instance; D: *DRIVEWAY* instance; RT: *ROAD-TERMINATOR* instance. (Instances of *RECTANGULAR-HOUSE*, *VISIBLE-ROAD-PIECE*, *RECTANGLE*, and *PICTURE-BOUNDARY* are not shown.)

object instances are shown in Fig. 5.35. Table 5.10 summarizes the statistics of the analysis results of Fig. 5.33.

This image includes several houses with dark roofs. Initially, all such houses were not extracted because the initial segmentation process assumes that houses have light roofs. As shown in Fig. 5.35c, three *HOUSE* instances with dark roofs were successfully detected by the top-down analysis; the lower threshold operator was applied to extract the dark

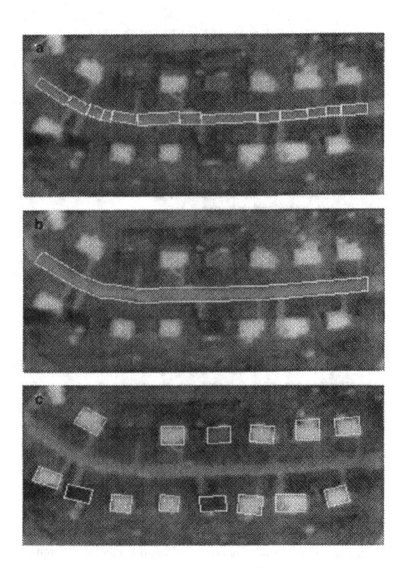

Figure 5.35. Iconic descriptions of recognized object instances in Fig. 5.33. (a) *ROAD-PIECE* instances. (b) *ROAD* instances. (c) *HOUSE* instances. (d) *HOUSE-GROUP* instances. (e) *DRIVEWAY* instances. (f) *ROAD-TERMINATOR* instances.

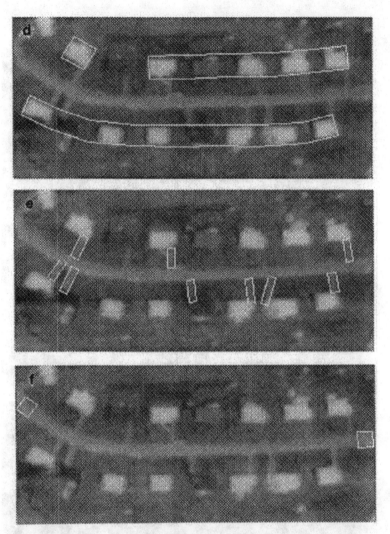

Figure 5.35. (*continued*)

rectangles after the other operators, assuming light ones, failed. Five
HOUSE instances do not have clear driveways. In this example, since no
RECTANGLE instances were incorrectly recognized as *ROAD-PIECE* instances,
no incorrect interpretation networks were generated.

In summary, although the system as implemented is simple, the
experimental results have demonstrated that it can stably analyze fairly
complex aerial images. In particular, extensive use of top-down analysis

Table 5.10. Statistics of the Analysis Result of Fig. 5.33

Object class	Initial No.	Final No.	Correct No.	Incorrect No.
RECTANGLE	22	40		
HOUSE	11	14	14	0
HOUSE-GROUP	0	3	3	0
ROAD-PIECE	9	13	13	0
ROAD	0	1	1	0
DRIVEWAY	0	9	9	0
ROAD-TERMINATOR	0	2	0	0

enables the stable recognition of objects with nonprominent features such as driveways. Considering that no color or texture properties were used in the system, the analysis results obtained in the experiments are very satisfactory. With such additional information, SIGMA will be able to analyze more complex aerial images like those used in our former system (Naga1980).

Finally, we could not measure computation time meaningfully because the system contains much auxiliary code to monitor intermediate analysis results.

Chapter 6

Conclusion

We have described the SIGMA image understanding system: motivations and design principles in Chapter 1, reasoning scheme and knowledge representation in Chapter 2, practical algorithms in Chapters 3 and 4, and experimental results in Chapter 5.

We started the design of SIGMA in 1983 and have since been studying various problems in image understanding. The work described in this book is a summary of our efforts during these years. During the study, we have uncovered many more technical problems than we have solved by the proposed methods. In this sense, the reasoning and knowledge representation methods described here are just a step toward true knowledge-based image understanding.

In this chapter, we first summarize three major achievements in SIGMA and discuss technical problems that must be solved in order to make further progress: evidence accumulation for spatial reasoning, object-oriented knowledge representation, and knowledge-based image segmentation. Then, we discuss areas for future work: three-dimensional object recognition and parallel distributed image understanding.

6.1. SUMMARY AND DISCUSSION

Our fundamental principle in designing SIGMA was that image understanding is not a passive interpretive reasoning process but an active constructive reasoning activity. In order to model such constructive reasoning, we proposed the use of evidence accumulation for spatial reasoning and object-oriented knowledge representation. We also identified the imperfection of segmentation as an essential problem in image understanding; the performance of many of the image understanding systems so far developed suffers considerably from segmentation errors. To address this problem, we proposed the extensive use of top-down goal-directed segmentation and a knowledge-based segmentation expert.

In what follows, we summarize the characteristics of these proposed methods and discuss remaining technical problems.

6.1.1. Evidence Accumulation for Spatial Reasoning

The most outstanding characteristics of our reasoning method are active hypothesis generation and the accumulation of partial evidence from different sources. This active reasoning method enables us to realize versatile spatial reasoning:

1. *Integration of bottom-up and top-down analyses.* Bottom-up and top-down analyses are integrated into a uniform reasoning process: one of these analysis processes is activated dynamically depending on the situation under analysis. Thus, *ad hoc* control knowledge about when to activate the top-down analysis is not required.

2. *Compensation for incomplete information.* The process of active hypothesis generation compensates for both insufficient input information and the imperfection of segmentation: a description of a partially occluded object can be completed by introducing hypotheses for the occluded parts, and hypotheses are used as goals to guide the top-down segmentation for missing objects.

3. *Reliable analysis.* The accumulation of partial evidence increases the reliability of the analysis. The integration of information from different sources is a general principle to realize reliable image understanding.

As discussed in Section 2.3.5, our reasoning method can be modeled as so-called hypothesis-based reasoning, in which semantics and the roles of (logical) hypotheses in the reasoning process are formally defined in terms of the first-order predicate calculus. Although hypothesis-based reasoning provides a formal foundation for our reasoning method, it should be extended further to manage erroneous observations (i.e., segmentation errors). In Section 2.3.6, we proposed a criterion to select meaningful observations.

We believe that it is crucial for image understanding systems to have well-defined logical/mathematical foundations; without such foundations, we cannot characterize reasoning schemes clearly nor compare them with others. Although it is very difficult to describe complex reasoning schemes completely in terms of formal logic and/or mathematics, their capabilities and limitations are clarified by studying their formal foundations. In fact, we have pinpointed the following theoretical and practical problems whose solution would improve the performance of our reasoning method.

6.1.1.1. Consistency Examination

In SIGMA, consistency between pieces of evidence (i.e., hypotheses and object instances) implies that they denote the same entity. As discussed in Section 3.4.3, the current consistency examination method used by SLDM is not complete. There are three major reasons for this:

1. *Consistency of knowledge sources.* Since the knowledge for spatial reasoning is distributed over many object classes, it may not be totally consistent. The system has no way to examine the consistency among such distributed knowledge sources. Note that if concrete geometric object models like those used in computer-aided design are given as knowledge, we can assure the consistency between pieces of evidence based on the models. In such cases, however, the system can recognize only instances of such fixed models. That is, consistency can be guaranteed at the sacrifice of flexibility. In other words, the richer and more flexible knowledge sources are, the harder their consistency examination becomes. How to maintain the consistency in a knowledge base is a major problem in artificial intelligence.

2. *Disjunctive knowledge.* In SIGMA, although alternative hypotheses are generated based on disjunctive knowledge, there is no explicit information representing the disjunction that only one of them can be verified. Metarules used in SIGMA are not capable of managing disjunctive knowledge; when one among several alternative hypotheses is verified, a metarule is activated to eliminate the others. Since the verified hypothesis may not be correct (due to segmentation errors), we should keep all alternative hypotheses and associate them with explicit descriptions representing the disjunction. Such descriptions could be used in consistency examination to avoid internal conflicts in interpretation networks.

3. *Segmentation errors.* It is almost impossible to prevent erroneous image features from being extracted from complex natural images. They sometimes happen to be recognized as objects, which makes consistency examination difficult. Since the question of how to cope with such erroneous information is beyond the scope of formal logic, we need to introduce an extralogical criterion based on which the correctness of information is evaluated. Although the maximal set of explainable observations proposed in Section 2.3.6 is a possible criterion, we should study its characteristics and computational algorithms.

One promising way of improving our consistency examination method is to incorporate the Assumption-Based Truth Maintenance System (ATMS) proposed by de Kleer (de K1986). Specifically, the Situation Lattice Database Manager should be augmented to perform flexible truth maintenance as well as consistency examination. Since we construct multiple interpretation networks which may be mutually conflicting, ATMS is best suited to maintain consistency and manage conflicting relations among such alternative interpretations. The incorporation of ATMS would make our consistency examination more rigorous.

Last, as discussed in Section 2.3.6, SIGMA does not use any negative information, and only positive (i.e., consistent) pieces of evidence are accumulated. The problem with internal conflicts discussed in Section 3.4.3 is ascribed partly to this inability to reason based on negative evidence. Since negative/positive evidence is determined by the consistency examination process, reasoning based on negative evidence should also be studied in the context of consistency examination.

6.1.1.2. Utilization of Relational Constraints

A hypothesis is a description of an expected (undetected) object. Thus, it should contain the same types of information as an object instance: attributes (i.e., intrinsic properties) and relational descriptions. In the current implementation of SIGMA, while attributes of a hypothesis are described by a set of constraints (i.e., inequalities), no relational constraints are associated with hypotheses. Therefore, after a solution to a hypothesis is provided, its source object instance sometimes has to perform complicated processing to verify relational constraints.

For example, as shown in Fig. 3.14e, a *ROAD* instance has to examine the continuity between a detected *ROAD-PIECE* instance and itself, because that *ROAD-PIECE* instance may not be adjacent to the *ROAD* instance.

Moreover, the ROAD instance needs the knowledge to perform further reasoning when the detected ROAD-PIECE instance is not adjacent to it. In Fig. 3.14e, for example, the ROAD instance directly activates the Model Selection Expert to fill the gap between the ROAD-PIECE instance and itself. If we could associate relational constraints with hypotheses (e.g., if a ROAD-PIECE hypothesis included the relational constraint that it must be adjacent to its source ROAD instance), then such complicated processings by object instances would not be required. Moreover, using relational constraints associated with hypotheses, the reliability of the top-down image segmentation would be increased.

However, the association of relational constraints with hypotheses will introduce further complications into the consistency examination; the consistency between relational constraints associated with different hypotheses must be examined. The difficulty of this consistency examination can be well understood by considering the situation shown in Fig. 3.14d: a pair of ROAD-PIECE hypotheses are partially overlapping and associated with adjacency constraints to different ROAD instances. Moreover, the Low-Level Vision Expert (LLVE) must also be augmented so that it can extract image features which satisfy specified geometric relations with other image features. Possible augmentations to LLVE will be discussed in Section 6.1.3.

6.1.1.3. Reliability Computation

To compute and associate a numerical reliability value with each piece of evidence is useful for several reasoning processes:

1. Image features and object instances with low reliability values can be eliminated as errors. This mechanism is a useful practical method to complement the incompleteness of the consistency examination.

2. Reasoning based on reliability values improves the robustness of the analysis. Since many pieces of evidence are integrated, small errors in feature extraction and measurement can be moderated during the reliability computation. Moreover, even partially conflicting pieces of evidence can be combined through reliability computation. The introduction of reliability values is especially useful to realize *graceful degradation of performance*; the performance of reasoning based on formal logic is drastically decreased by a minor modification of knowledge (axioms).

3. As discussed in Section 3.5.2, the entire reasoning process can be controlled based on reliability values, which is useful to realize

efficient analysis: the analysis can be focused on the most reliable partial interpretation, leaving the others unanalyzed.

Although the performance of SIGMA will be improved by introducing numerical reliability computation, we need to augment ordinary probability models (i.e., Bayesian and Dempster and Shafer models); as discussed in Section 3.5.1, while they assume a fixed set of entities for probability computation, many new entities are dynamically inserted in SIGMA. Currently, the first author is studying the logical foundations of the Dempster and Shafer model with a view to applying it to image understanding.

Another practical problem in reliability computation is how we can represent the reliability of evidence by a numerical value. Most systems have used ad hoc reliability assignment rules. For example, the longer a line segment, the larger is the reliability value assigned. Since a piece of evidence has many different properties (i.e., attributes and relations), we need well-defined foundations to transform its property values into a numerical reliability value.

6.1.1.4. Computation Cost

SIGMA actively generates many hypotheses to complement insufficient input information and the imperfection segmentation. To generate multiple descriptions of an object (i.e., hypotheses and instances) is the key mechanism to realize such versatile reasoning. From a computation cost point of view, however, we need large memory space and much computation time to manage such duplicated descriptions. The considerable requirement for memory space and computation time constitutes a common practical problem in both evidential reasoning and hypothesis-based reasoning.

Although we used several mechanisms to reduce the number of pieces for evidence in the database—for example, hypothesis generation by hypotheses is prohibited and verified/refuted hypotheses are immediately removed—they are not sufficient. In order to reduce the computation cost, new mechanisms should be introduced. The global focus of attention mechanism discussed in Section 3.5.2 will help to increase efficiency. Moreover, since we prevent hypotheses from generating new hypotheses, the computational complexity itself is not great (see Section 3.4.1); therefore, parallel processing will be useful to improve the speed of the analysis. A discussion of parallel processing will be offered in Section 6.2.

6.1.2. Object-Oriented Knowledge Representation

In the reasoning scheme of SIGMA, object instances are not static data structures to record properties of recognized objects but active reasoning agents which perform spatial reasoning about their surrounding environments. We used object-oriented knowledge representation to realize such distributed image understanding.

The knowledge required for an object instance to perform reasoning is represented by production rules, which are stored in its corresponding object class: all instances of the same object class share the same set of rules. In this sense, our knowledge representation can be considered as a mixture of frame and production systems.

Moreover, in order to realize flexible spatial reasoning, rules in SIGMA are augmented to improve upon ordinary IF-THEN rules: ⟨hypothesis⟩ parts are additionally introduced into the rules. ⟨Hypothesis⟩ parts play a crucial role in coordinating reasoning activities by individual object instances. That is, the execution of an ⟨action⟩ part is delayed until its corresponding hypothesis has been processed (i.e., verified/refuted). Recall that a hypothesis is processed when a situation including it is resolved and that a situation usually consists of multiple hypotheses. Hence, when a situation is resolved, multiple delayed actions are executed simultaneously. In other words, the execution of actions is synchronized through their corresponding hypotheses. In this sense, ⟨hypothesis⟩ parts can be considered as a mechanism to realize synchronization among independent reasoning agents (i.e., object instances). In Section 6.2, the synchronization mechanism will be discussed in the context of parallel processing.

As noted in Section 6.1.1.1, it is difficult to examine the overall consistency among all sets of production rules separately stored in object classes. However, our object-oriented knowledge representation has several advantages.

6.1.2.1. Modularity

It is widely accepted that a modular system organization is very useful in the development of large software systems. Modularity facilitates both incremental system development and maintenance. Object-oriented computation is a powerful scheme to realize modular system organization.

As advocated in Mins1975, it is natural to organize mutually related pieces of knowledge into a common module (i.e., object class in SIGMA). Such modular knowledge organization facilitates the implemenation and modification of knowledge. Moreover, the inheritance mechanism allows multiple modules to share common pieces of knowledge. This knowledge sharing is useful not only to reduce the required memory space but also to simplify maintenance; if the same pieces of knowledge were separately stored in different modules, when pieces of knowledge in one module were modified, we would have to apply the same modification to those in the others.

6.1.2.2. Partitioning of Production Rules

In ordinary production systems, a set of rules is stored in one place and the reasoning engine examines the entire set to select applicable ones at every recognize-and-act cycle. In SIGMA, on the other hand, the rule set is partitioned into subsets, each of which is stored separately in an object class. Such partitioning of rules has the following advantages and disadvantages:

- *Advantages*. It increases the efficiency of the rule selection as well as the modularity of the knowledge representation. Since each subset of rules can be examined independently, rule selection and application can be done in parallel.
- *Disadvantages*. Recall that a pair of rules which are used to perform spatial reasoning based on the same *PART-OF*/spatial relation are stored separately in different object classes. The separation of such mutually related rules makes consistency examination difficult. We need a new mechanism to maintain consistency between separately stored rules.

6.1.2.3. Semantic Attachments of Relational Links

In our knowledge organization, five types of relational links between object classes are used: *A-KIND-OF*, *APPEARANCE-OF*, in-conflict-with, *PART-OF*, and spatial relation. A major problem in knowledge representation is how to define the semantics of such relational links. For example, in object-oriented programming languages, the semantics of *A-KIND-OF* links is defined by the property inheritance mechanism. In ACRONYM (Broo1981), the meaning of *PART-OF* links is defined in terms of geometric transformations.

In SIGMA, while the semantics of in-conflict-with links is implicitly defined by procedures used in consistency examination, two types of semantic attachments are used to define the semantics of the other relational links:

1. *Rules for spatial reasoning.* PART-OF and spatial relations are used for spatial reasoning. Their semantics is defined by production rules stored in object classes. Since we can represent various control conditions and analysis procedures by rules, very flexible spatial reasoning can be performed based on these relational links.

2. *Transformation functions.* As discussed in Section 3.4.2, A-KIND-OF and PART-OF links are used to transform a description of one object class to that of another in consistency examination. We associate a pair of functions with each link to realize such transformation. That is, the semantics of A-KIND-OF and PART-OF links is defined by associated transformation functions.

   Originally, APPEARANCE-OF links were introduced to represent the mapping between the scene and image domains. In the current system, however, their semantics is defined by transformation functions similar to those for A-KIND-OF links. We need to introduce new semantic attachments for APPEARANCE-OF links in order to really implement the mapping. This point will be discussed in Section 6.2 in the context of three-dimensional object recognition.

6.1.3. Knowledge-Based Image Segmentation

Since SIGMA extensively uses top-down image segmentation to detect missing objects, it is crucial for the system to have a capable image segmentation module. LLVE was developed to realize reliable image segmentation. It has its own knowledge about image processing techniques and reasons about the most effective image analysis process to detect image features specified in a given goal. Moreover, LLVE performs trial-and-error image analysis automatically to select appropriate operators and to adjust parameters. Using LLVE, SIGMA can find many objects which could not be detected by uniform initial segmentation. We believe that the introduction of such knowledge-based segmentation experts into image understanding systems greatly increases their performance.

The LLVE developed in this study is just a prototype and could be improved in various ways. In Section 4.5, we proposed heterogeneous compositions of image processing operators and discussed extensions of the knowledge representation scheme of LLVE. With such extensions, the capability of LLVE will be increased considerably. The following are suggestions for further extensions.

6.1.3.1. Extraction of Complex Image Features

Image features extractable by the current LLVE are confined to very primitive ones such as linear segments and rectangles. It cannot extract complex image features with internal structures, e.g., a light region with a narrow bottleneck and two dark compact holes. In order to realize the extraction of complex image features, we need to introduce a new module on top of LLVE, one which would decompose a complex image feature into simpler ones and guide LLVE to detect each of the simple image features. Note that this new module should have reasoning capability to guide LLVE; when some simple image features have been detected, it has to reason about which missing image feature should be specified as the next goal to LLVE. This reasoning process shares much with the process of spatial reasoning by GRE.

6.1.3.2. Reasoning Based on Relational Constraints

As discussed in Section 6.1.1.2, when relational constraints are associated with hypotheses, a goal given to LLVE comes to include constraints on geometric relations. "Find a rectangle which is adjacent to ⟨Rectangle123⟩" is an example of such a goal. A straightforward way to use this geometric constraint is to first extract all possible rectangles and then to select ones which satisfy the constraint (i.e., adjacency to ⟨Rectangle123⟩). In this method, however, the image analysis process itself does not use the specified geometric constraint at all and hence the performance of LLVE is not improved. How to best use the given geometric constraints during the image analysis is an interesting topic for future research.

6.1.3.3. Smooth Interface between Signal and Symbolic Data

LLVE can be considered as an interface between signal and symbolic data processing; input signal data is converted into symbolic data by LLVE. This conversion involves approximation and decision: approximate a connected component of pixels by a polygon, recognize (decide) a region as a rectangle, and so on. Since image features and their properties are in general fuzzy, the conversion should be performed by taking such fuzziness into account. In order to realize a smooth interface between signal and symbolic data, the knowledge representation and reasoning method of LLVE should be augmented with fuzzy predicates and/or probabilistic reasoning.

6.2. AREAS FOR FUTURE WORK

While the previous section dealt with suggestions for augmentations and improvements to the current implementation SIGMA, the following are research topics for future work.

6.2.1. Three-Dimensional Object Recognition

Although the framework of SIGMA is general enough for image understanding of the three-dimensional (3D) scenes, many technical problems must be solved before SIGMA can be applied to 3D object recognition.

6.2.1.1. Representation of 3D Object Models

In SIGMA, *APPEARANCE-OF* links are prepared to describe 3D object models: a 3D object model is described by a set of 2D appearances from different viewpoints. This appearance-based model representation is useful to recognize 3D objects from a 2D image (Mins1975, Goad1983, Ikeu1987). Given a 3D geometric model, however, it is very hard to generate mutually all its possible 2D appearances and describe their mutual relations. Thus, we have to develop a *model compiler* which produces a set of meaningful 2D appearances automatically (Goad1983, Ikeu1987). Aspect graphs (Koen1979) are a typical appearance-based model representation, and currently efficient algorithms to generate an aspect graph from a 3D object model are being studied in the field of computer vision (Watt1988).

6.2.1.2. Appearance Model Selection

In analyzing 3D scenes, appearance model selection becomes very important because it involves the determination of the mapping between 3D object models and 2D image features. That is, in appearance-based model representation, the selection of a specific 2D appearance implies the determination of the viewpoint of a 3D object. Since in general an image feature matches many appearances of 3D objects, flexible and efficient reasoning capabilities are required for appearance model selection. Ikeuchi (Ikeu1987) organized 2D appearances of a 3D object into a decision tree, based on which reasoning about viewing angles is conducted. That is, a strategy for appearance model selection is represented by the decision tree. In order to realize efficient appearance model selection, we need such control knowledge to guide the reasoning. In other words, rich control information should be associated with *APPEARANCE-OF* links by their semantic attachments.

6.2.1.3. Utilization of Stereo, Motion, and Range Images

Although it is not impossible to recognize 3D objects from a single 2D image, the amount of input information is very limited. By using stereo, motion, and range images, we can directly measure the 3D information of the scene, which would substantially enrich the input information. As is well known, however, the measurement of 3D information from stereo and motion images involves a difficult problem (i.e., the correspondence problem), which requires extensive study in itself.

6.2.2. Parallel Distributed Image Understanding

The reasoning scheme used in SIGMA includes much potential parallelism. By implementing SIGMA on a multiprocessor system, we could make use of the following types of parallel processing.

6.2.2.1. Parallel Application of Rules

As discussed in Section 6.1.2.2, production rules in SIGMA are partitioned into disjoint subsets. Since for each object instance the rule (sub)set in its corresponding object class is examined and applied to generate hypotheses independently of the others, the rule application to each object instance could be executed in parallel. In other words, we can regard each object instance as a parallel process executed on a different process, which generates hypotheses in parallel based on its own rules. We call this parallel reasoning scheme *parallel distributed image understanding* (see also Section 6.2.2.3). Note that although the rule sets in an object class are shared by all of its instances, no mechanism for exclusive access is required since the rules are not modified by the instances (i.e., they are read-only).

6.2.2.2. Parallel Consistency Examination

Usually multiple hypotheses are inserted into the Iconic/Symbolic Database at the same time (i.e., at the beginning of the same interpretation cycle). The consistency examination between newly generated hypotheses and pieces of evidence stored in the database can be done in parallel: for each new hypothesis, generate a parallel process to examine its consistency with the other pieces of evidence. In this case, such parallel processes must communicate with each other to examine the consistency with other new hypotheses. Moreover, since the situation lattice is shared and modified by the parallel processes, mechanisms for synchronization and exclusive access are required.

It is also possible for each parallel process to spawn parallel subprocesses, each of which examines the consistency between a new hypothesis and a piece of evidence in the database. This pairwise consistency examination can be carried out completely in parallel since it involves no data modification.

Note that these types of parallel processing are done internally in GRE. That is, in our parallel distributed image understanding model, GRE as well as object instances is implemented on a group of processors.

6.2.2.3. Parallel Reasoning by Object Instances

Since there are many situations to be resolved at each interpretation cycle, their resolution can be performed in parallel. Since the resolution of a situation activates multiple delayed actions, the parallel resolution of multiple situations means that many object instances are activated in parallel to execute delayed actions. That is, in our parallel distributed image understanding model, each object instance (i.e., parallel process) performs reasoning (i.e., executes actions) as well as generates hypotheses (i.e., applies rules) in parallel.

6.2.2.4. Synchronization

As noted in Section 6.1.2, parallel reasoning activities by object instances are synchronized through hypotheses they have generated. Suppose object instances **a** and **b** generated hypotheses $f(\mathbf{a})$ and $g(\mathbf{b})$, respectively, which are accumulated to form a situation. Reasoning activities based on (i.e., execution of ⟨action⟩ parts of) the rules which generated $f(\mathbf{a})$ and $g(\mathbf{b})$ are delayed until that situation is resolved. That is, the resolution of that situation requires the synchronized activation of parallel reasoning by **a** and **b**. The action list can be regarded as a queue to realize the synchronization. In this sense, GRE can be considered as a synchronization mechanism among parallel processes. This synchronization mechanism is very similar to the rendezvous in the Ada programming language (Ben1982), which is a useful synchronization/communication mechanism among concurrent processes.

6.2.2.5. OR Parallelism

By OR parallelism we mean parallel processing based on disjunctive knowledge. When there are two possibilities to pursue, say A and B, a pair of parallel processes are generated to examine A and B respectively. In SIGMA, alternative hypotheses are generated based on disjunctive knowledge, which can be analyzed in parallel. In order to realize such parallel processing, we need a new synchronization mechanism to coordinate parallel processes generated based on OR parallelism: when one of them finds a correct answer successfully, the other must be forced to be terminated.

In our parallel distributed image understanding model described above, much parallelism can be extracted. However, there are many difficult problems in its implementation:

1. Since each object instance generates hypotheses and modifies its internal data independently of the others, it is very hard to maintain consistency among such distributed stored information. This is an essential problem in distributed problem solving.
2. The above descriptions concern a model of parallel processing. We now face the task of designing software and hardware architectures to implement them: determining how we can describe behaviors of complicated parallel processes, and what hardware architectures are suitable to implement parallel distributed image understanding systems.

In conclusion, to realize a parallel distributed image understanding system is a challenging research topic, one which will lead us to a new paradigm of image understanding as well as to the realization of very fast image understanding systems. The work described in this book can be considered as a step toward parallel distributed image understanding.

References

Abbreviations used in references:

AAAI: National Conference on Artificial Intelligence
CVPR: Computer Vision and Pattern Recognition
GE: Geoscience and Remote Sensing
ICPR: International Conference on Pattern Recognition
IJCAI: International Joint Conference on Artificial Intelligence
IPS: Information Processing Society
PAMI: Pattern Analysis and Machine Intelligence
SMC: IEEE Systems, Machine and Cybernetics

Agin1979   G. J. Agin: Knowledge-Based Detection and Classification of Vehicles and Other Objects in Aerial Images, Proceedings of the DARPA Image Understanding Workshop, Palo Alto, CA, pp. 66–71, April 1979.

Aho1972   A. Aho and J. Ullman: *Theory of Parsing, Translation, and Compiling*, Prentice-Hall, Englewood Cliffs, NJ, 1972.

Andr1988   K. M. Andress and A. C. Kak: Evidence Accumulation and Flow of Control in a Hierarchical Spatial Reasoning System, *AI Magazine*, Vol. 9, No. 2, pp. 75–94, 1988.

Ball1976   D. H. Ballard: *Hierarchic Recognition of Tumors in Chest Radiographs*, Birkhauser-Verlag, Basel, 1976.

Ball1981   D. H. Ballard: Generalizing the Hough Transform to Detect Arbitrary Shapes, *Pattern Recognition*, Vol. 13, No. 2, pp. 111–112, 1981.

Ball1982   D. H. Ballard and C. M. Brown: *Computer Vision*, Prentice-Hall, Englewood Cliffs, NJ, 1982.

Barn1981   J. A. Barnett: Computational Methods for a Mathematical Theory of
            Evidence, Proceedings of the 7th IJCAI, Vancouver, BC, Canada, pp.
            868–875, August 1981.

Barr1971   H. G., Barrow and R. J. Popplestone: Relational Descriptions in Picture
            Processing, *Machine Intelligence,* Vol. 6, pp. 377–396, 1971.

Barr1976   H. G. Barrow and J. M. Tenenbaum: MSYS: A System for Reasoning
            about Scenes, Technical Note 121, SRI International, Menlo Park, CA,
            April 1976.

Barr1978   H. G. Barrow and J. M. Tenenbaum: Recovering Intrinsic Scene Charac-
            teristics from Images, in *Computer Vision Systems* (A. R. Hanson and E.
            M. Riseman, eds.), pp. 3–26, Academic Press, New York, 1978.

Barr1982   A. Barr and E. A. Feigenbaum (eds.): *The Handbook of Artificial
            Intelligence,* Chapter X, William Kaufmann, Los Altos, CA, 1982.

Ben1982   M. Ben-Ari: *Principles of Concurrent Programming,* Prentice-Hall, Engle-
            wood Cliffs, NJ, 1982.

Bent1979   J. L. Bentley and J. H. Friedman: Data Structures for Range Searching,
            ACM Computing Surveys, Vol. 11, No. 4, pp. 397–409, 1979.

Binf1982   T. O. Binford: Survey of Model-Based Image Analysis Systems,
            International Journal of Robotic Research, Vol. 1, No. 1, pp. 18–64, 1982.

Boll1976   R. C. Bolles: Verification Vision within a Programmable Assembly System,
            Memo AIM-295, Stanford Artificial Intelligence Laboratory, Stanford, CA,
            1976.

Broo1981   R. A. Brooks: Symbolic Reasoning about 3-D Models and 2-D Images,
            Artificial Intelligence, Vol. 17, pp. 285–348, 1981.

Chan1973   C. L. Chang and R. C. T. Lee: *Symbolic Logic and Mechanical Theorem
            Proving,* Academic Press, New York, 1973.

Cloc1981   W. F. Clocksin and C. S. Mellish: *Programming in Prolog,* Springer-
            Verlag, New York, 1981.

Cox1981   P. T. Cox and T. Pietrzykowski: Deduction Plans: A Basis for Intelligent
            Backtracking, *IEEE Transactions,* Vol. PAMI-3, No. 1, pp. 52–65, 1981.

Cox1986   P. T. Cox and T. Pietrzykowski: *Causes for Events: Their Computation and
            Applications,* Lecture Notes on Computer Science, No. 230, pp. 608–621,
            Springer-Verlag, Berlin, 1986.

Davi1976   L. S. Davis and A. Rosenfeld: Applications of Relaxation Labeling 2:
            Spring-Loaded Template Matching, Technical Report 440, Computer Sci-
            ence Center, University of Maryland, College Park, 1976.

Davi1978   L. S. Davis and A. Rosenfeld: Hierarchical Relaxation for Waveform
            Parsing, in *Computer Vision Systems* (A. R. Hanson and E. M. Riseman,
            eds.), pp. 101–109, Academic Press, New York, 1978.

de K1986   J. de Kleer: An Assumption-Based TMS, *Artificial Intelligence,* Vol. 28, pp.
            127–224, 1986.

Doyl1979   J. Doyle: A Truth Maintenance System, *Artificial Intelligence,* Vol. 28, pp.
            127–224, 1986.

Doyl1979   J. Doyle: A Truth Maintenance System, *Artificial Intelligence,* Vol. 12, pp.
            231–272, 1979.

Duda1973   R. O. Duda and P. E. Hart: *Pattern Classification and Scene Analysis,* John
            Wiley & Sons, New York, 1973.

Duda1976   R. O. Duda, P. Hart, and N. J. Nilsson: Subjective Bayesian Methods for
            Rule-Based Inference Systems, in *Proceedings of the National Computer
            Conference,* pp. 1075–1082, AFIPS Press, Arlington, VA, 1976.

Faug1982   O. D. Faugeras: Relaxation Labeling and Evidence Gathering, Proceedings of the 6th ICPR, Munich, FRG, pp. 405–412, 1982.

Feke1984   G. Fekete and L. S. Davis: Property Spheres: A New Representation for 3D Object Recognition, Proceedings of the Workshop on Computer Vision, Annapolis, MD, pp. 192–201, 1984.

Fing1985   J. J. Finger and M. R. Genesereth: RESIDUE: A Deductive Approach to Design Synthesis, Technical Report CS-85-1035, Stanford University, Stanford, CA, 1985.

Fisc1973   M. A. Fischler and R. A. Elschlager: The Representation and Matching of Pictorial Structures, *IEEE Transactions*, Vol. C-22, No. 1, pp. 67–92, 1973.

Fish1987   R. Fisher: SMS: A Suggestive Modelling System for Object Recognition, *Image and Vision Computing*, Vol. 5, No. 2, pp. 98–104, 1987.

Forg1982   C. L. Forgy: Rete: A Fast Algorithm for the Many Pattern/Many Object Pattern Match Problem, *Artificial Intelligence*, Vol. 19, pp. 17–37, 1982.

Free1974   H. Freeman: Computer Processing of Line Drawings, *Computing Surveys*, Vol. 6, pp. 57–98, 1974.

Free1975   J. Freeman: The Modelling of Spatial Relations, *Computer Graphics and Image Processing*, Vol. 4, pp. 156–171, 1975.

Fuku1988   K. Fukushima: A Neutral Network for Visual Pattern Recognition, *IEEE Computer*, Vol. 21, No. 3, pp. 65–75, 1988.

Garv1976   T. D. Garvey: Perceptual Strategies for Purposive Vision, Technical Note 117, SRI International, Menlo Park, CA, 1976.

Garv1981   T. D. Garvey, J. D. Lowrance, and M. A. Fishler: An Inference Technique for Integrating Knowledge from Disparate Sources, Proceedings of the 7th IJCAI, Vancouver, BC, Canada, pp. 319–325, August 1981.

Garv1987   T. D. Garvey: Evidential Reasoning for Geographic Evaluation for Helicopter Route Planning, *IEEE Transactions*, Vol. GE-25, No. 3, pp. 294–304, 1987.

Gene1987   M. R. Genesereth and N. J. Nilsson: *Logical Foundations of Artificial Intelligence*, Morgan Kaufmann, Los Altos, CA, 1987.

Goad1983   C. Goad: Special Purpose Automatic Programming for 3D Model-Based Vision, Proceedings of the DARPA on Image Understanding Workshop, Arlington, VA, pp. 94–104, 1983

Gold1983   A. Goldberg and D. Robson: *Smalltalk-80: The Language and Its Implementation*, Addison-Wesley, Reading, MA, 1983.

Greg1970   R. L. Gregory: *The Intelligent Eye*, McGraw-Hill, New York, 1970.

Haar1982   R. L. Haar: Sketching: Estimating Object Positions from Relational Descriptions, *Computer Graphics and Image Processing*, Vol. 19, pp. 227–247, 1982.

Haas1987   L. J. de Haas: Automatic Programming of Machine Vision Systems, Proceedings of the 10th IJCAI, pp. 790–792, 1987.

Hans1978   A. R. Hanson and E. M. Riseman: VISIONS: A Computer System for Interpreting Scenes, in *Computer Vision Systems* (A. R. Hanson and E. M. Riseman, eds.), pp. 303–333, Academic Press, New York, 1978.

Hara1979   R. M. Haralick and L. G. Shapiro: The Consistent Labeling Problem: Part I, *IEEE Transactions.*, Vol. PAMI-1, pp. 173–184, 1979.

Hara1981   R. M. Haralick and L. Watson: A Facet Model for Image Data, *Computer Graphics and Image Processing*, Vol. 15, pp. 113–129, 1981.

Hase1987   J. Hasegawa, H. Kubota, and J. Toriwaki: IMPRESS: A System for Image Processing Procedure Construction Based on Sample-Figure Presentation,

Transactions, IEICE of Japan, Vol. J70-D, No. 11, pp. 2147–2153, 1987 [in Japanese].

Hase1988   J. Hasegawa, H. Kubota, A. Takasa, and J. Toriwaki: Consolidation of Image Processing Procedures in the Image Processing Expert System IMPRESS, in Special Issue on Expert Systems for Image Processing, *Journal of the IPS of Japan,* Vol. 29, No. 2, pp. 126–133, 1988 [in Japanese].

Have1983   W. Havens and A. Mackworth: Representing Knowledge of the Visual World, *IEEE Computer,* Vol. 16, No. 10, pp. 90–96, 1983.

Haye1983   F. Hayes-Roth, D. A. Waterman, and D. B. Lenat (eds.): *Building Expert Systems,* Addison-Wesley, Reading, MA, 1983.

Herm1986   M. Herman and T. Kanade: Incremental Reconstruction of 3D Scenes from Multiple, Complex Images, *Artificial Intelligence,* Vol. 30, pp. 289–341, 1986.

Hild1979   E. Hildreth and D. Marr: Theory of Edge Detection, Memo 518, MIT Artificial Intelligence Laboratory, Cambridge, MA, April 1979.

Hild1983   E. C. Hildreth: Computations Underlying the Measurement of Visual Motion, *Artificial Intelligence,* Vol. 22, pp. 1–27, 1983.

Humm1983   R. A. Hummel and S. W. Zucker: On the Foundations of Relaxation Labeling, *IEEE Transactions,* Vol. PAMI-5, No. 3, pp. 267–287, 1987.

Hwan1984a   V. Hwang: Evidence Accumulation for Spatial Reasoning in Aerial Image Understanding, Ph.D. Thesis, University of Maryland, College Park, 1984.

Hwan1984b   V. Hwang, T. Matsuyama, L. S. Davis, and A. Rosenfeld: Evidence Accumulation for Spatial Reasoning in Aerial Image Understanding, Proceedings of the 7th ICPR, Montreal, Canada, pp. 394–397, July 1984.

Hwan1986   V. Hwang, L. S. Davis, and T. Matsuyama: Hypothesis Integration in Image Understanding Systems, *Computer Vision, Graphics, and Image Processing,* Vol. 36, pp. 321–371, 1986.

Ikeu1987   K. Ikeuchi: Generating an Interpretation Tree from a CAD Model for 3D-Object Recognition in Bin-Picking, *International Journal of Computer Vision,* Vol. 1, No. 2, pp. 145–165, 1987.

Kell1971   M. D. Kelly: Edge Detection in Pictures by Computer Using Planning, in *Machine Intelligence* (B. Meltzer and D. Michie, eds.), Vol. 6, pp. 377–396, Edinburgh University Press, Edinburgh, 1971.

Koen1979   J. J. Koenderik and A. J. Van Doorn: The Internal Representation of Solid Shape with Respect to Vision, *Biological Cybernetics,* Vol. 32, pp. 211–216, 1979.

Lloy1984   J. W. Lloyd: *Foundations of Logic Programming,* Springer-Verlag, New York, 1984.

Lowe1985a   D. G. Lowe: *Perceptual Organization and Visual Recognition,* Kluwer, Boston, 1985.

Lowe1985b   D. G. Lowe: Visual Recognition from Spatial Correspondence and Perceptual Organization, Proceedings of the 9th IJCAI, Los Angeles, CA, pp. 953–959, August 1985.

Lowe1987   D. G. Lowe: The Viewpoint Consistency Constraint, *International Journal of Computer Vision,* Vol. 1, No. 1, pp. 57–72, 1987.

Lowr1982   J. D. Lowrance: Dependency-Graph Models of Evidential Support, COINS Technical Report 82–26, University of Massachusetts, Amherst, 1982.

McDe1980a   J. McDermott: R1: An Expert in the Computer Systems Domain, Proceedings of the National AI Conference, Stanford, CA, pp. 269–271, August 1980.

McDe1980b   D. McDermott: A Theory of Metric Spatial Inference, Proceedings of the National AI Conference, Stanford, CA, pp. 246–248, August 1980.

McKe1983   D. M. McKeown: MAPS: The Organization of a Spatial Database System Using Imagery, Terrain, and Map Data, Proceedings of the Workshop on Image Understanding, Arlington, VA, pp. 105–127, June 1983.

McKe1985   D. M. McKeown, W. A. Harvey, and J. McDermott: Rule-Based Interpretation of Aerial Imagery, *IEEE Transactions,* Vol. PAMI-7, No. 5, pp. 570–585, 1985.

Marr1975   D. Marr: Early Processing of Visual Information, Memo 340, MIT Artificial Intelligence Laboratory, Cambridge, MA, 1975.

Marr1978   D. Marr: Representing Visual Information, in *Computer Vision Systems* (A. R. Hanson and E. M. Riseman, eds.), Academic Press, New York, pp. 61–80, 1978.

Marr1982   D. Marr: *Vision,* Freeman, 1982.

Mats1984a   T. Matsuyama and T. Y. Phillips: Digital Realization of the Labeled Voronoi Diagram and Its Application to Closed Boundary Detection, Proceedings of the 7th ICPR, pp. 478–480, 1984.

Mats1984b   T. Matsuyama, L. V. Hao, and M. Nagao: A File Organization for Geographic Information Systems Based on Spatial Proximity, *Computer Vision, Graphics, and Image Processing,* Vol. 26, pp. 303–318, 1984.

Mats1984c   T. Matsuyama, H. Arita, and M. Nagao: Structural Matching of Line Drawings Using the Geometric Relationship between Line Segments, *Computer Vision, Graphics, and Image Processing,* Vol. 27, pp. 177–194, 1984.

Mats1985   T. Matsuyama and V. Hwang: SIGMA: A Framework for Image Understanding, Proceedings of the 9th IJCAI, pp. 908–915, 1985.

Mats1986   T. Matsuyama and M. Ozaki: LLVE: An Expert System for Top-Down Image Segmentation, *Journal of the IPS of Japan,* Vol. 27, pp. 191–204, 1986 [in Japanese].

Mats1987   T. Matsuyama: Knowledge-Based Aerial Image Understanding Systems and Expert Systems for Image Processing, *IEEE Transactions,* Vol. GE-25, No. 3, pp. 305–316, 1987.

Mats1988   T. Matsuyama, N. Murayama, and T. Ito: On Representation of Image Analysis Strategies, in Special Issue on Expert Systems for Image Processing, *Journal of the IPS of Japan,* Vol. 29, No. 2, pp. 169–177, 1988 [in Japanese].

Mend1964   E. Mendelson: *Introduction to Mathematical Logic,* Van Nostrand, Princeton, NJ, 1964.

Milg1979   D. L. Milgram: Region Extraction Using Convergent Evidence, *Computer Graphics and Image Processing,* Vol. 11, pp. 1–12, 1979.

Mins1975   M. Minsky: A Framework for Representing Knowledge, in *Psychology of Computer Vision* (P. H. Winston, ed.), McGraw-Hill, New York, 1975.

Mont1971   U. Montanari: On the Optimal Detection of Curves in Noisy Pictures, *Communications of the ACM,* Vol. 14, No. 5, pp. 335–345, 1971.

Nadi1984   A. Nadif and M. D. Levine: Low Level Image Segmentation: An Expert System, *IEEE Transactions.,* Vol. PAMI-6, No. 5, pp. 555–577, 1984.

Naga1980   M. Nagao and T. Matsuyama: *A Structural Analysis of Complex Aerial Photographs,* Plenum Press, New York, 1980.

Naga1984   M. Nagao: Control Structures in Pattern Analysis, *Pattern Recognition,* Vol. 17, pp. 45–56, 1984.

Naga1988   H. Nagahashi and M. Nakatsuyama: A Study of Expert Systems in Image
             Processing, in Special Issue on Expert Systems for Image Processing,
             Journal of the IPS of Japan, Vol. 29, No. 2, pp. 88–95, 1988 [in Japanese].
Neva1980   R. Nevatia and K. Babu: Linear Feature Extraction, *Computer Graphics
             and Image Processing*, Vol. 13, pp. 257–269, 1980.
Newe1978   A. N. Newell, N. J. McDermott, and J. Moore: The Efficiency of Certain
             Production System Implementations, in *Pattern-Directed Inference Systems*
             (D. A. Waterman and F. Hayes-Roth, eds.), Academic Press, New York,
             1978.
Nils1980   N. J. Nilsson: *Principles of Artificial Intelligence*, Tioga, Palo Alto, CA, 1980.
Ohta1980   Y. Ohta: A Region-Oriented Image-Analysis System by Computer, Ph.D.
             Thesis, Kyoto University, 1980.
Ohta1985   Y. Ohta and T. Kanade: Stereo by Intra- and Inter-Scanline Search Using
             Dynamic Programming, *IEEE Transactions.*, Vol. PAMI-7, No. 2, pp.
             139–154, 1985.
Ohts1979   N. Ohtsu: A Threshold Selection Method from Gray-Level Histogram,
             IEEE Transactions., Vol. SMC-9, pp. 62–66, 1979.
Parm1980   C. C. Parma, A. M. Hanson, and E. M. Riseman: Experiments in
             Schema-Driven Interpretation of a Natural Scene, COINS Technical Report
             80-10, University of Massachusetts, Amherst, 1980.
Pool1985   D. Poole: On the Comparison of Theories: Preferring the Most Specific
             Explanation, Proceedings of the 9th IJCAI, Los Angeles, CA, pp. 144–145,
             August 1985.
Pool1987   D. Poole, R. Aleliunas, and R. Goebel: Theorist: A Logical Reasoning
             System for Defaults and Diagnosis, in *The Knowledge Frontier: Essays in
             the Representation of Knowledge* (N. J. Cercone and G. McCalla, eds.),
             Springer-Verlag, New York, 1987.
Prep1972   F. P. Preparata and S. R. Ray: An Approach to Artificial Non-Symbolic
             Cognition, *Information Science*, Vol. 4, pp. 65–86, 1972.
Pric1982   K. E. Price: Symbolic Matching of Images and Scene Models, Proceedings
             of the DARPA Workshop on Image Understanding, Palo Alto, CA, pp.
             299–308, September 1982.
Quam1978   L. H. Quam: Road Tracking and Anomaly Detection in Aerial Imagery,
             Proceedings of the DARPA Image Understanding Workshop, Palo Alto,
             CA, pp. 51–55, May 1978.
Redd1979   R. Reddy: Note, Proceedings of the Workshop on Control Structures and
             Knowledge Representation for Image and Speech Understanding, Univer-
             sity of Maryland, College Park, April 3–4, 1979.
Reit1980   R. Reiter: Equality and Domain Closure in First-Order Databases, *Journal
             of the ACM*, Vol. 27, NO. 2, pp. 235–249, 1980.
Reit1987   R. Reiter and A. K. Mackworth: The Logic of Depiction, Technical Report
             RCVB-TR-87-18, University of Toronto, 1987.
Robe1963   L. G. Roberts: Machine Perception of Three-Dimensional Solids, in *Optical
             and Electro-Optical Information Processing* (J. T. Tippett et al., eds.), pp.
             159–197, MIT Press, Cambridge, MA, 1963.
Rose1976   A. Rosenfeld, R. A. Hummal, and S. W. Zucker: Scene Labeling by
             Relaxation Operations, *IEEE Transactions.*, Vol. SMC-6, No. 6, pp.
             420–433, 1976.
Rose1978   D. A. Rosenthal: An Inquiry Driven Computer Vision System Based on
             Visual and Conceptual Hierarchies, Ph.D. Thesis, University of Penn-
             sylvania, Philadelphia, 1978.

Rose1982   A. Rosenfeld and A. C. Kak: *Digital Picture Processing*, Vols. 1 and 2, Academic Press, New York, 1982.

Rubi1980   S. M. Rubin: Natural Scene Recognition Using Locus Search, *Computer Graphics and Image Processing*, Vol. 14, No. 4, pp. 298–333, 1980.

Russ1979   D. M. Russell: Where Do I Look Now? Proceedings of Pattern Recognition and Image Processing Conference, Long Beach, CA, pp. 175–183, 1979.

Saka1985   K. Sakaue and H. Tamura: Automatic Generation of Image Processing Programs, Proceedings of CVPR, San Francisco, CA, pp. 189–192, 1985.

Self1982   P. G. Selfridge: Reasoning about Success and Failure in Aerial Image Understanding, Ph.D. Thesis TR103, University of Rochester, 1982.

Shaf1975   G. Shafer: *A Mathematical Theory of Evidence*, Princeton University Press, Princeton, NJ, 1975.

Shaf1983   S. A. Shafer and T. Kanade: The Theory of Straight Homogeneous Generalized Cylinders, Proceedings of the DARPA Image Understanding Workshop, Arlington, VA, pp. 210–218, June 1983.

Sham1975   M. I. Shamos and D. Hoey: Closest Point Problems, Proceedings of the 16th Annual Symposium on Foundations of Computer Science, pp. 131–162, 1975.

Shor1976   E. H. Shortliffe and H. Edward: *Computer-Based Medical Consultations: MYCIN*, American Elsevier, New York, 1976.

Smit1878   R. G. Smith and R. Davis: Distributed Problem Solving: The Contract Net Approach, Report STAN-CS-78-667, Computer Science Department, Stanford University, Stanford, CA, 1978.

Spec1980   Special Issue on Non-Monotonic Reasoning, *Artificial Intelligence*, Vol. 13, 1980.

Spec1984   Special Issue on Qualitative Reasoning, *Artificial Intelligence*, Vol. 24, 1984.

Spec1988   Special Issue on Expert Systems for Image Processing, *Journal of the IPS of Japan*, Vol. 29, No. 2, 1988 [in Japanese].

Sued1986   N. Sueda and H. Hoshi: An Expert System for Designing Image Analysis Programs Based on Software Packages, in *Expert Systems—Theory and Application*, Nikkei Business Publications, Tokyo, pp. 135–154, 1986 [in Japanese].

Tamu1983   H. Tamura, S. Sakane, F. Tomita, N. Yokoya, M. Kanoko, and K. Sakane: Design and Implementation of SPIDER—A Transportable Image Processing Software Package, *Computer Vision, Graphics, and Image Processing*, Vol. 23, pp. 273–294, 1983.

Tamu1988   H. Tamura, H. Sato, K. Sakane, and F. Kubo: DIA-Expert System and Its Knowledge Representation Scheme, in Special Issue on Expert Systems for Image Processing, *Journal of the IPS of Japan*, Vol. 29, No. 2, pp. 199–207, 1988 [in Japanese].

Tani1975   S. Tanimoto and T. Pavlidis: A Hierarchical Data Structure for Picture Processing, *Computer Graphics and Image Processing*, Vol. 4, No. 2, pp. 104–119, 1975.

Tene1977   J. M. Tenenbaum and H. G. Barrow: Experiments in Interpretation-Guided Segmentation, *Artificial Intelligence*, Vol. 8, pp. 241–274, 1977.

Tori1982   J. Toriwaki, K. Mase, Y. Yashima, and T. Fukumura: Modified Voronoi Diagrams and Relative Neighbors on a Digital Picture and Their Applications to Tissue Image Analysis, Proceedings of the 1st International Symposium on Medical Imaging and Image Interpretation, pp. 362–367, 1982.

Tori1987   T. Toriu, H. Iwase, and M. Yoshida: An Expert System for Image
               Processing, *Fujistu Scientific Technical Journal,* Vol. 23, No. 2, pp.
               111–118, 1987.

Tsot1981   J. K. Tsotsos: Event Recognition: An Application to Left Ventricular
               Performance, Proceedings of the 7th IJCAI, Vancouver, BC, Canada, pp.
               900–907, August 1981.

Tsot1988   J. K. Tsotsos: A "Complexity Level" Analysis of Immediate Vision,
               International Journal of Computer Vision, Vol. 1, No. 4, pp. 303–320, 1988.

Walt1975   D. Waltz: Understanding Line Drawings of Scenes with Shadows, in
               Psychology of Computer Vision (P. H. Winston, ed.), McGraw-Hill, New
               York, 1975.

Watt1988   N. A. Watts: Calculating the Principal Views of a Polyhedron, Proceedings
               of the 9th ICPR, pp. 316–322, 1988.

Wein1980   D. Weinreb and D. Moon: Flavors: Message Passing in the LISP Machine,
               Memo 602, MIT Artificial Intelligence Laboratory, 1980.

Wesl1982   L. P. Wesley and A. R. Hanson: The Use of an Evidential-Based Model for
               Representing Knowledge and Reasoning about Images in the VISIONS
               System, Proceedings of the IEEE Workshop on Computer Vision, Rindge,
               NH, pp. 14–25, 1982.

Wins1975   P. H. Winston: Learning Structural Descriptions from Examples, in
               Psychology of Computer Vision (P. H. Winston, ed.), McGraw-Hill, New
               York, 1975.

Witk1983   A. P. Witkin: Scale-Space Filtering, Proceedings of the 8th IJCAI,
               Karlsruhe, FRG, pp. 1019–1022, August 1988.

Wood1978   R. J. Woodham: Photometric Stereo: A Reflectance Map Technique for
               Determining Surface Orientation from Image Intensity, Proceedings of the
               22nd International Symposium Society of Photo-Optical Instrumentation
               Engineering, pp. 136–143, 1978.

Yaki1973   Y. Yakimovsky and J. A. Feldman: A Semantic-Based Decision Theory
               Region Analyzer, Proceedings of the 3rd IJCAI, Stanford, CA, pp.
               580–588, 1973.

Zade1973   L. A. Zadeh: Fuzzy Sets, *Information and Control,* Vol. 8, pp. 338–353,
               1965.

Zahn1974   C. T. Zahn: An Algorithm for Noisy Template Matching, Proceedings of
               IFIP 74, pp. 727–732, 1974.

Zuck1975   S. W. Zucker, A. Rosenfeld, and L. S. Davis: General Purpose Models:
               Expectations about the Unexpected, Proceedings of the 4th IJCAI, Tbilisi,
               Georgia, USSR, pp. 716–721, September, 3–8, 1975.

Zuck1976   S. W. Zucker: Relaxation Labeling and the Reduction of Local Ambi-
               guities, Technical Report 451, Computer Science Department, University of
               Maryland, College Park, 1976.

Index